Bernhard Kienzler and Juliet S. Swanson

# Microbial Effects in the Context of Past German Safety Cases

Karlsruhe Institute of Technology
**KIT SCIENTIFIC REPORTS 7744**

# Microbial Effects in the Context of Past German Safety Cases


by
Bernhard Kienzler
Karlsruhe Institute for Technology (KIT) | Institute for Nuclear Waste Disposal (INE)

Juliet S. Swanson
Los Alamos National Laboratory


KIT Scientific Publishing

Report-Nr. KIT-SR 7744

Funding for the Los Alamos Actinide Chemistry and Repository Science Program contribution to this report was provided by the U.S. Department of Energy, Carlsbad Field Office.
This report was approved for public release with unlimited distribution as LA-UR-17-27144.

**Impressum**

KIT Scientific Publishing

Karlsruher Institut für Technologie (KIT)
KIT Scientific Publishing
Straße am Forum 2
D-76131 Karlsruhe

KIT Scientific Publishing is a registered trademark
of Karlsruhe Institute of Technology.
Reprint using the book cover is not allowed.

www.ksp.kit.edu



Print on Demand 2017 – Gedruckt auf FSC-zertifiziertem Papier

ISSN 1869-9669
ISBN 978-3-7315-0712-3
DOI 10.5445/KSP/1000073559

# Abstract

Microbes can be found virtually everywhere including deep geological disposals for radioactive wastes. Internationally, microbial effects are considered in the scope of many Safety Cases, in Germany these issues received minor attention. In order to provide for a supplement to the ongoing investigations on microbial effects in underground disposals, this review describes the waste treatment procedures and organic inventory in German facilities, the microbial processes to be considered, such as aerobic and anaerobic processes, microbial sulfate reduction, the pH range and chaotropic effects on microbial activity. The microbial effects in repositories are discussed on the basis of effects depending on resident or introduced microorganisms and the environmental conditions causing waste form degradation.

Microbial effects on barriers of a repository and on the migration and retention of radionuclides are also presented. The microbial population of the cap rock above the Gorleben salt dome and the effect of microorganisms on the retention of technetium and selenium are shown. In the preliminary safety analysis of Gorleben (vSG), microbial sulfate reduction was analyzed which might affect the mechanical properties of a salt repository.

Finally, the results of the EU projects Funmig and ReCosy are referred and some conclusions are drawn.

# Kurzfassung


Mikrobielle Effekte im Zusammenhang früherer deutscher Sicherheitsbetrachtungen

Mikroben werden praktisch überall gefunden, einschließlich in tiefen geologischen Endlagern für radioaktive Abfälle. International werden mikrobielle Effekte im Rahmen vieler Sicherheitsnachweisen berücksichtigt, in Deutschland wurden diese Fragen nur selten direkt behandelt. Um die derzeit laufenden Untersuchungen zu mikrobiellen Wirkungen in unterirdischen Endlagern zu ergänzen, beschreibt diese Übersicht die Abfallbehandlungsverfahren und die organischen Inventare in deutschen Anlagen, die relevanten mikrobiellen Prozesse, wie z. B. aerobe und anaerobe Prozesse, mikrobielle Sulfatreduktion, den pH-Bereich und die chaotropen Effekte hinsichtlich ihres Einflusses auf die mikrobielle Aktivität. Die mikrobiellen Effekte in den Endlagern werden auf der Grundlage von residenten oder eingeführten Mikroorganismen geführt und hängen von den Umgebungsbedingungen ab.

Auch mikrobielle Effekte auf die Barrieren eines Endlagers und auf die Migration und Rückhaltung von Radionukliden werden vorgestellt. Die mikrobielle Population der überlagernden Gesteinsschichten über dem Gorleben-Salzstock und die Wirkung von Mikroorganismen bezüglich der Rückhaltung von Technetium und Selen werden aufgezeigt. In der vorläufigen Sicherheitsanalyse Gorleben (vSG) wurde die mikrobielle Sulfatreduktion analysiert, die die mechanischen Eigenschaften eines Endlagers im Steinsalz beeinflussen könnte.

Abschließend wird auf die Ergebnisse der EU-Projekte Funmig und ReCosy verwiesen und es werden einige Schlussfolgerungen gezogen.

# Content

# Figures

# Tables

# Abbreviations

| | |
|---|---|
| $a_W$ | Chemical water activity |
| Bac. | Bacterium |
| BSR | Microbial sulfate reduction |
| CEC | Cation exchange capacity |
| CFU | Colony Forming Units |
| DMS | Deutsche Sammlung von Mikroorganismen (German Collection of Microorganisms) |
| ERAM | Endlager für radioaktive Abfälle Morsleben (Morsleben Repository for low-level Radioactive Wastes) |
| Funmig | EURATOM EC 6$^{th}$ Framework Program Integrated Project "Fundamental Processes of Radionuclide Migration" (2005-2008) |
| HZDR | Helmholtz Zentrum Dresden Rossendorf |
| ILW | Intermediate level radioactive waste |
| KfK | Kernforschungszentrum Karlsruhe, today KIT |
| Konrad | Licensed disposal for low radioactive wastes in the Konrad iron ore mine at Salzgitter |
| LLW | Low level radioactive waste / LAW |
| -Log $m_{H+}$ | Proton concentration in a solution (pH) |
| MIND | EURATOM EC Horizon 2020 Collaborative Project "Microbiology in Nuclear Waste Disposal" (2015-2018) |
| NF-Pro | EURATOM EC 6$^{th}$ Framework Program Integrated Project "Understanding and Physical and Numerical Modelling of the Key Processes in the Near Field and their Coupling for Different Host Rocks and Repository Strategies" (2004-2007) |
| Pamina | EURATOM EC 6$^{th}$ Framework Program Integrated Project Performance Assessment Methodologies in Application to Guide the Development of the Safety Case" (2006-2009) |
| PE | Polyethylene |
| PVC | Polyvinylchloride |

| | |
|---|---|
| ReCosy | EURATOM EC 7th Framework Program Collaborative Project "Redox Phenomena Controlling Systems" (2008-2011) |
| TIC | Total inorganic carbon |
| TSR | Thermochemical sulfate reduction |
| URL | Underground laboratory |
| vSG | Vorläufige Sicherheitsanalyse Gorleben (vSG): Preliminary Safety Analysis of Gorleben |

# Preface

From time to time, questions concerning the effect of microbial activities on the safety of underground nuclear disposals arise. In the INIS databe of the IAEA, 247 publications are registered having the terms "microbial" and "disposal" in the titles. The first publications appeared since 1971. Presently, about 2 to 8 new publications per year are found in INIS. The highest number of publications appeared in 2012 with 17 new publications, partly as a result of the conference series on "clays in natural and engineered barriers for radioactive waste confinement" organized by European waste management organizations (WMO).

Since 2015, the European Commission has funded the Horizon 2020 project "Microbiology in Nuclear Waste Disposal (MIND)[a]". The MIND project brings together 15 European groups working on the impact of microbial processes on safety cases for geological repositories, focusing on key questions posed by waste management organizations throughout Europe.

In Deliverable 1.1 of MIND entitled "A Review of Anthropogenic Organic Wastes and Their Degradation Behavior"[1], the inventories of organic components in various European nuclear waste disposal concepts are presented. Data are available from Belgium, the Czech Republic, Finland, France, Spain, Sweden, the Netherlands and United Kingdom. The German partner in the project (HZDR) did not provide information on inventories.

KIT is not a partner of the MIND project. However, KIT-INE possesses information on investigations and considerations of microbial effects within the German disposal programs since 1967. For this reason, a supplement to the MIND review is provided: "Microbial Effects in the Context of Past German Safety Cases".

---

[a] Euratom grant agreement 661880 (2015-2019).

# 1 Introduction

The knowledge of microbial processes affecting repository safety and performance is incomplete because current safety assessment concepts either ignore microbial processes, or adopt a simplified approach. In December 2016, an article was published in the German journal “Spektrum der Wissenschaften” raising the question if microbes can live in a nuclear waste repository and if they could cause problems in the disposal safety or if they could increase disposal safety [2]. The article deals with microbes under aerobic conditions prevailing in the Swedish Äspö underground research laboratory and in the former Königstein uranium mining facility in Germany. It is undoubtedly correct that microorganisms exist in deep mines and are present in the waste forms and in the barrier materials.

A variety of anthropogenic polymers and organic wastes (including plastic, textile materials, paper and ion exchange materials among others) might be present in nuclear waste disposals. In addition, bitumen is used in several European countries as an encapsulant for nuclear waste and organic additives are present in cement used in the different barriers. These organic materials provide an energy and carbon source that has the potential to fuel microbiological processes in a repository. An estimation of the potential inventory of organic matter has been elaborated for the Swedish spent fuel disposal [3] and for the planned French disposal in argillite [4]. In Germany, the effects of microorganisms on disposal safety has not been profoundly considered [5]. For this reason, the available information on organics in radioactive waste materials in German waste disposals are summarized including a literature overview on the typical degradation processes and some investigations performed in Germany. This overview considers mainly low-level or intermediate-level wastes, as they may contain organics which can be substrates for microorganisms, metallic components, which can undergo corrosion processes caused by microorganisms, and the presence of cemented waste forms, which increase the pH.

The realization that microbes could exist in the environment of a nuclear waste repository posed a series of questions [6]:

1. Do microorganisms occur naturally in deep geological formations?
2. Can introduced microorganisms survive after repository closure?
   If the answer to these is yes,
3. What effect have the environmental conditions of a nuclear waste repository on resident or introduced microorganisms?
4. What effect will microorganisms have on a repository in terms of structural integrity of the engineered barriers, or radionuclide release and subsequent migration?

For bituminized waste forms microbial degradation studies were performed and the results are published specifically in regard to the Waste Isolation Pilot Plant (WIPP). A summary of some reports on microbial degradation of bitumen is given in this study. Other topics covered by this summary are attributed to the microbial degradation of paper, plastics and textiles which is an important issue of the Asse II safety case. For the Konrad LLW disposal in Germany, studies on the microbial degradation of compacted mixed wastes were performed. These studies were not published but presented in the minutes of the panel dealing with low- and intermediate waste forms („Arbeitskreis LAW/MAW-Produkte")[b]. The objective of the panel was to provide methods for characterization and control procedures for waste forms [7].

Further grey literature information is provided for microbial attack of cement, investigation of the microbiological populations in the cap rocks above the Gorleben salt dome as well as the analyses of microbial effects on mobilization of radionuclides. Considerations concerning the sulfate reduction processes are described, which were a topic of the preliminary safety analysis of the disposal in the Gorleben salt dome.

[b] In the minutes of the 16th meeting of the panel (June 10, 1986 at Kernforschungszentrum Karlsruhe), a study was presented dealing with the stability of low- and intermediate level wastes against microbial attacks.

# 2 Waste treatment and organic inventory

Since the 1980's, the treatment of radioactive wastes has reached a high technical standard. According to the nuclear industry in Germany (Arbeitskreis Abfallmanagement des VGB PowerTech e.V., www.vgb.org/abfallmanagement.html, 2004) the treatment procedures distinguish between solid and liquid wastes, wastes suited for incineration, and other categories. The different treatment procedures are shown in Fig. 1. In principle, the same methods were used for reprocessing and laboratory wastes. For waste streams consisting of paper, plastics and textiles both incineration and compaction procedures are possible, according to the specific radioactivity and other properties.

The huge amount of L/ILW from reprocessing required additional pretreatment steps than the ones specified in Fig 1. The methods applied for reprocessing wastes include evaporation, ion exchange and chemical flocculation and were effective in terms of decontamination factors, volume reduction and reliability. For the solidification of the waste concentrates arising from the decontamination of the liquid effluents several appropriate methods were available, such as the incorporation into cement, bitumen and plastics. Fixation in cement was in principle the simplest and the most cost effective method. A main disadvantage of cementation was the doubling of the waste volume. Bituminization was considered superior to cementation. In comparison with cementation a volume reduction by a factor of 2-6 was achieved.

A series of organic solidification matrices have been applied, especially for conditioning wastes from nuclear power plants [8]:

- Bitumen
- Urea-formaldehyde
- Organoplastics (polyesters, epoxy-resins and polystyrene)
- Vinylesters

The properties of the resulting waste forms have been studied, including mechanical stability, leaching behavior and radiolytical degradation. As these materials were not thermodynamically stable, the long-term degradation processes were also considered.

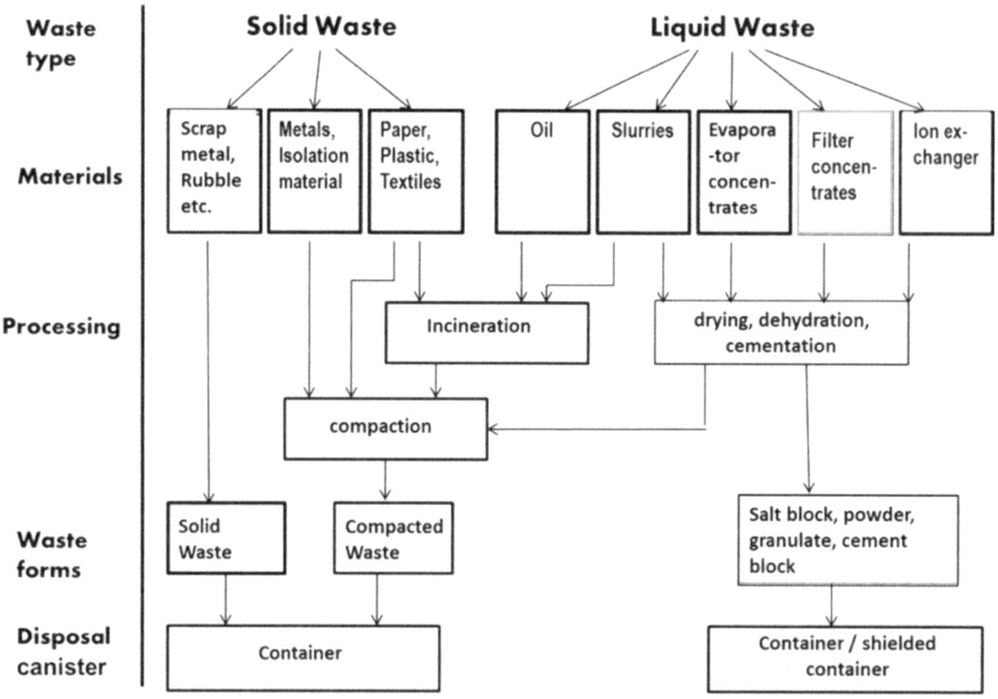


Fig. 1 Optimized treatment procedures for solid and liquid radioactive wastes (www.vgb.org/abfallmanagement.html).

Since 1967, about 126 000 low-level waste canisters have been disposed of in the Asse II mine until termination of the disposal activities in 1978. The organic content of the Asse II mine was estimated as follows: 2% of the wastes were solidified in bitumen matrices, 0.8% in various polymers, and 2% consist of absorbent materials. 8% of the wastes contained cellulose, paper or wood for the most part lightly compacted in steel drums. In total, about 1200 Mg of cellulose material have been disposed of in the Asse mine [9].

In the scope of the preliminary safety analysis for the Gorleben nuclear waste disposal (vSG), an estimation was compiled of the inventories of organic waste forms in the different German interim storage facilities [10]. The result was that about 136 Mg of bituminized waste forms and 49 Mg wastes solidified in plastics exist and need to be disposed of.

# 3 Microbial processes

Karsten Pedersen, MICANS made following statement in his article "The deep biosphere" published in GFF, the Journal of the Geological Society of Sweden [11]: "*Microbes can be found virtually everywhere in the tree of life. They constitute the absolutely dominant diversity of life on our planet and the total amount of carbon in these intra-terrestrial organisms may equal that of all terrestrial and marine plants. Biochemically, much of this diversity is contradictory to multi-cellular life whose diversity is largely morphological. The enormous biochemical diversity among the microbes explains their huge adaptability to almost any environment on the planet where temperature allows life. Some Archaea likes it very hot. For instance, the optimum temperature for growth of the genus Pyrolobus is 105°C and it survives in temperatures of up to 113°C.*" He also stated that "*microbes certainly must obey the universal laws saying that matter and energy cannot be destroyed or formed, just transformed. But they are utterly sophisticated lawyers that understand to use every single, constructive combination of those laws for making a good living of all or almost nothing*".

In the deep biosphere, the microbial activities are in equilibrium with their environment. Different environments will support different microbial communities. For example, fractures in a crystalline rock-based repository may allow the influx of gases ($H_2$) that can be used as energy sources for microorganisms or may introduce groundwater containing other possible organic or inorganic electron donors. In contrast, the impermeability of salt (e.g., WIPP) will likely prevent this type of biosphere. Iron redox chemistry, in a repository, will exert a significant control on microbial activity.

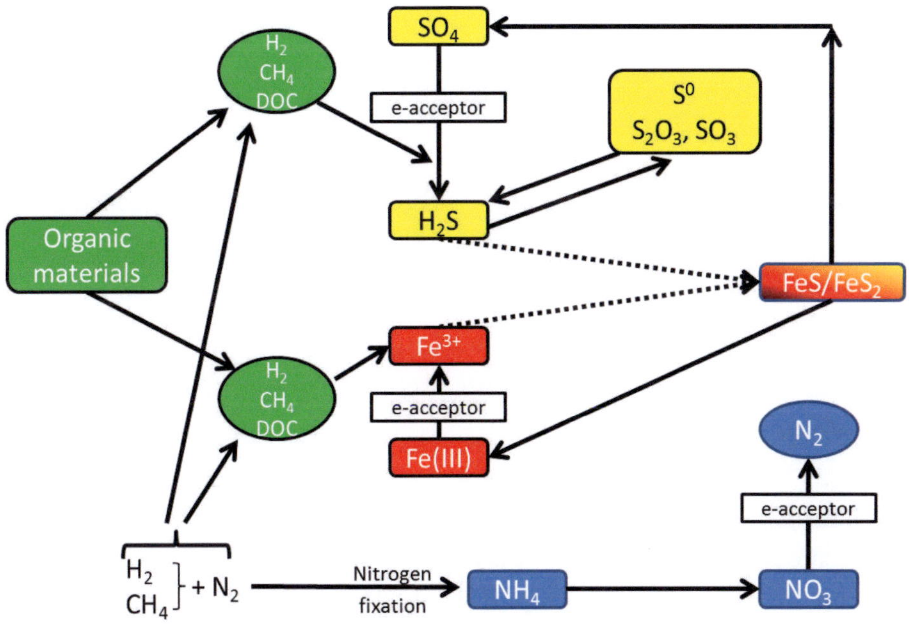


Fig. 2 Schematic view of microbial processes in the deep underground.

In the following chapter, several of the processes shown in Fig. 2 are described. It is obvious that the construction and operation of a nuclear waste disposal changes some of the environmental conditions for the microorganisms. Oxygen becomes available for some time, organic materials might be constituents of the wastes, or the backfill materials and hydrogen will be produced by corrosion processes of metals in the disposal.

Microbial metabolisms can be arranged according to some basic principles which are shown in Fig. 3 [12]:

1. How the organism obtains energy for living and growing:
   chemotrophic – energy is obtained from external chemical compounds
   phototrophic – energy is obtained from light
2. How the organism obtains reducing equivalents used either in energy conservation or in biosynthetic reactions:
   lithotrophic – reducing equivalents are obtained from inorganic compounds
   organotrophic – reducing equivalents are obtained from organic compounds
3. How the organism obtains carbon for synthesizing cell mass:
   autotrophic – carbon is obtained from carbon dioxide ($CO_2$)
   heterotrophic – carbon is obtained from organic compounds
   mixotrophic – carbon is obtained from both organic compounds
   and by fixing carbon dioxide

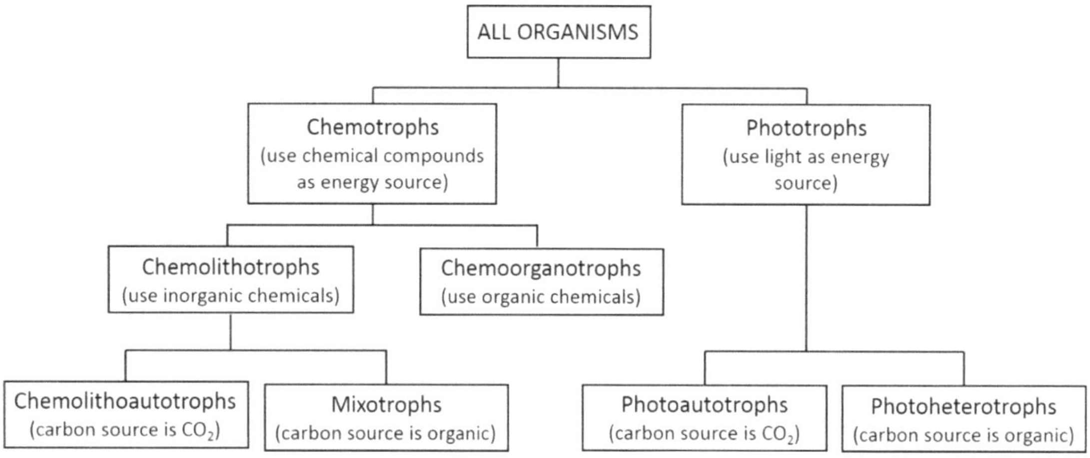


Fig. 3 Microbial metabolisms [12].

## 3.1 Aerobic processes

Aerobic dissimilation: Aerobic organisms during respiration use oxygen as a terminal electron acceptor. The complete degradation of glucose $(CH_2O)_n$ by microbes might be considered as reversal of the photosynthesis reaction. Reaction (1) represents an aerobic dissimilation process.

$C_6H_{12}O_6 + 6O_2 + 6H_2O \rightarrow 12H_2O + 6CO_2$ ($\Delta G_0$ =−2880 kJ/mol glucose) (1)

The energy release of this reaction is used for adenosine triphosphate (ATP) or acetate production in the cell. In the case of anaerobic dissimilation, instead of oxygen other compounds are used as electron acceptors, such as nitrate ($NO_3^-$), $Fe^{3+}$, and sulfate ($SO_4^{2-}$). The respiratory strategies of the microorganisms include

1. direct enzymatic reduction at the outer membrane,
2. electron shuttling pathways and
3. metal dissolution by exogenous or bacterially-produced organic ligands followed by reduction of soluble organic-metal compounds [13].

The last point is of specific relevance in the case of poorly soluble metals. The electron transport systems of gram-negative bacteria are generally described as inner membrane (IM)-associated electron and proton carriers that

- mediate electron transfer from primary donor to terminal electron acceptor and
- conserve energy released during electron transfer to the generation of ATP.

Metal reduction: Fe(III)- and Mn(IV)-respiring dissimilatory metal-reducing bacteria (DMRB) require interacting anaerobically with an electron acceptor in crystalline or amorphous form which cannot contact the IM-localized electron transport systems. In [13] it is assumed that Fe(III)- and Mn(IV)-respiring DMRB employ a variety of strategies for

1. direct enzymatic reduction of solid Fe(III) and Mn(IV) oxides via outer membrane (OM)-localized metal reductases,
2. a two-step, electron shuttling pathway in which electron shuttling compounds are first enzymatically reduced and subsequently chemically oxidized by the solid Fe(III) and Mn(IV) oxides in a second (abiotic) electron transfer reaction,
3. an analogous two-step reduction pathway involving endogenous, electron shuttling compounds, and
4. a twostep, Fe(III) dissolution-reduction pathway in which solid Fe(III) oxides are first dissolved by exogenous or bacterially-produced organic complexing ligands, followed by uptake and reduction of the soluble organic Fe(III) forms by periplasmic Fe(III) reductases.

Details of the electron transport and proton translocation processes of aerobic and anaerobic respiratory chains including the relevant enzymes, ADP and ATP as well as quinone pools etc. are described by DiChristina et al. [13].

With respect to the important iron reactions and the energy supply for microbial reactions relevant in a deep repository, data are provided by Kappler & Straub [14]. The authors provide a list of species which are able to affect the redox system controlled by iron. Table I shows some important microbial catalyzed iron redox transformations.

Tab. I Physiological groups of prokaryotes that catalyze iron redox transformations [14].

| Habitat | Electron donor | Electron acceptor | pH | Microbial metabolism | Representative strains |
|---|---|---|---|---|---|
| **Aerobic** | Fe(II) | $O_2$ | acidic | Fe(II) oxidation | *Thiobacillus ferrooxidans*<br>*Sulfobacillus acidophilus* |
| | Fe(II) | $O_2$ | neutral | Fe(II) oxidation | *Gallionella ferruginea*<br>*Leptothrix ochracea* |
| **Anoxic** | Fe(II) | $NO_3^-$ | neutral | $NO_3^-$-dependent Fe(II) oxidation | *Acidovorax sp. strain BrG1*<br>*Azospira strain PS is also perchlorate reducer* |
| | Fe(II) | $CO_2$ | neutral | Phototrophic Fe(II) oxidation | *Rhodobacter ferrooxidans strain SW2*<br>*Rhodovulum iodosum* |
| | Fe(II) | $ClO_3^-$ | neutral | Chlorate reduction | *Strain CKB [15]* |
| | Organic or inorganic compounds | Fe(III) | acidic | Fe(III) reduction | *Acidiphilium cryptum sp. JF-5*<br>*Thiobacillus thiooxidans* |
| | Organic or inorganic compounds | Fe(III) | neutral | Fe(III) reduction | *Geobacter metallireducens*<br>*Shewanella oneidensis* |

Even if the aerobic acidophilic Fe(II)-oxidizing microorganisms are irrelevant for deep underground disposal, some data should be mentioned: At acidic pH, the redox couple $Fe^{3+}/Fe^{2+}$ has a redox potential of +770 mV. At pH 2, only -33 kJ/mol iron is produced by oxidation with $O_2$, since the relevant redox potential of the redox couple $O_2/H_2O$ is +1106 mV. This difference allows the synthesis of 1 mol ATP.

Aerobic neutrophilic Fe(II)-oxidizing microorganisms: This physiological group of microorganisms uses $O_2$ as electron acceptor for enzymatic oxidation of Fe(II) at neutral pH. To gain energy for growth they have to compete with the chemical oxidation of Fe(II) by $O_2$.

## 3.2 Anaerobic processes

Anaerobic organisms do not need $O_2$ but use other electron acceptors. Anaerobic Fe(II)-oxidizing phototrophic bacteria which oxidize siderite to Fe(II) and $CO_2$ require light. These reactions are irrelevant for underground systems. Fermentation is a specific type of heterotrophic metabolism that uses organic carbon instead of oxygen as a terminal electron acceptor.

Anaerobic Fe(II)-oxidizing nitrate-reducing bacteria: Nitrate-dependent/iron-oxidizing bacterial isolates cannot be propagated for many generations as pure or mixed cultures, but apparently the original enrichments can, because they contain an autotrophic member that provides fixed carbon. Recently, some strains were isolated from the deep sea that oxidized Fe(II) with nitrate in the absence of an additional organic substrate. It was not clear whether these strains can actually grow with ferrous iron as the sole electron donor for several successive generations (2005).

Acidophilic Fe(III)-reducing microorganisms: Microbial reduction of ferric iron was known as a phenomenon for many decades before its (bio)geochemical relevance was recognized. At neutral pH, ferric iron is the dominant electron acceptor for the mineralization of carbon particularly in anoxic freshwater habitats.

Microbial iron cycling at neutral pH: Microbial iron cycling needs substrates, i.e., electron donors for Fe(III) reduction and electron acceptors for Fe(II) oxidation. Not all iron minerals are equally good substrates, as the redox potential of an iron redox couple determines whether the mineral is available as electron donor or acceptor. At pH 7, molecular oxygen and nitrate can accept electrons from ferrous iron, independently from the Fe(III) mineral produced. The situation is more complex with ferric iron oxides as electron acceptor. The oxidation of acetate ($CO_2$/acetate, $E_0$ = -290 mV) is energetically favorable only with specific iron oxides such as lepidocrocite or ferrihydrite. On the other hand, for the reduction of goethite, hematite or magnetite, electron donors with a lower redox potential are necessary, e.g., molecular hydrogen ($2H^+/H_2$, $E_0$ = -414 mV) or formate ($CO_2$/formate, $E_0$ = -432 mV). Hence, theoretically acetate can fuel microbial cycling of iron only if ferrihydrite or lepidocrocite is the product of microbial Fe(II) oxidation. Furthermore; it is essential that supplementary electron donors and acceptors can diffuse since ferric iron is barely soluble and thus rather immobile.

The standard redox potential of the $CO_2$/acetate, $CO_2$/methane, and sulfate/sulfide half-cell reactions approximates -290, -240 and -220 mV, respectively. Some relevant redox potentials for microbial reactions are listed in Tab. II.

Tab. II Redox half-reaction reduction potentials and free energies*

| Redox Pair (ox/red) | Electron donor | $E_O$ (volt) | $\Delta G_O$ (kJ/e$^-$) |
|---|---|---|---|
| $CO_2/CH_2O$ | carbohydrate | -0.43 | +41.5 |
| $CO_2/CH_3OH$ | methanol | -0.39 | +37.5 |
| $CO_2/CH_3COCOO^-$ | pyruvate | -0.37 | +35.8 |
| $CO_2/CH_2CHOHCOO^-$ | lactate | -0.34 | +32.9 |
| $CO_2/C_{16}H_{24}O_5N_4$ | protein | -0.33 | +32.2 |
| $CO_2/CH_3CH_2OH$ | ethanol | -0.33 | +31.8 |
| $CO_2/CH_3COO^-$ | acetate | -0.29 | +28.0 |
| $CO_2/C_2H5COO^-$ | propionate | -0.29 | +28.0 |
| $CO_2/C_8H_{16}O$ | oil and grease | -0.29 | +28.0 |
| $CO_2/CH_4$ | methane | -0.25 | +24.1 |
| $SO_4^{2-}/HS^-$ | sulfide | -0.22 | +20.9 |
| $NO_3^-/NH_4^+$ | ammonium | +0.36 | -34.7 |
| $NO_3^-/N_2$ | nitrogen | +0.75 | -72.4 |
| $Fe^{3+}/Fe^{2+}$ | ferrous iron | +0.77 | -74.3 |
| $O_2/H_2O$ | water | +0.82 | -79.1 |

* http://ceae.colorado.edu/~silverst/cven5534/REDOX%20HALF%20REACTIONS.pdf

# 3.3 Microbial sulfate reduction

## 3.3.1 Dissimilatory sulfate reduction

For sulfate reduction electron donors and energy are needed. The bacteria and archaea can oxidize organic compounds or molecular hydrogen ($H_2$) and reduce sulfate to hydrogen sulfide ($H_2S$). Some sulfate-reducing bacteria can also reduce other compounds such as nitrate or iron (or uranium). In this context, the term "dissimilatory" is used when hydrogen sulfide is produced and excreted in the anaerobic respiration process. Dissimilatory sulfate reduction uses sulfate as an electron acceptor. Sulfate must be activated by the enzyme ATP-sulfurylase, which uses ATP and sulfate to create adenosine 5'-phosphosulfate (APS). APS is subsequently reduced to sulfite ($SO_3^{2-}$) and AMP. Sulfite is then further reduced to sulfide ($S^{2-}$), while AMP is transferred into ADP using another molecule of ATP.

## 3.3.2 Anaerobic methane oxidation as respiratory process

From a thermodynamic point of view, sulfate-dependent methane oxidation is an exergonic reaction

$$CH_4 + SO_4^{2-} + 2H^+ \rightarrow CO_2 + H_2S + 2H_2O \quad (2)$$

which yields a $\Delta G_0$ of -18 kJ per mol under standard conditions. The concentrations of the chemical reaction partners *in situ* at the active layers are in the range of $10^3$ Pa of methane and about 2 to 3 mM each sulfate and free hydrogen sulfide. This amount of energy could feed only one single bacterium, provided that it is able to exploit this biological minimum energy quantum. The observations of the reversal of homoacetogenic fermentation e.g. by strain AOR allow to speculate that "reversed methanogenesis may be the key to this process. If the overall reaction is actually a syntrophic cooperation involving a methanogen running methane formation backward and a sulfate-reducing bacterium, it is obvious that only one of the partners could gain metabolically useful energy from the reaction and that the other one had to run this process only as a co-metabolic activity. This would explain why scientists have always failed to enrich for methane-oxidizing sulfate reducers in the past [1997], simply because one cannot enrich for bacteria on the basis of co-metabolic activities." Since Schink's review in 1997 [16] was published, anaerobic methane oxidizers have been identified by DNA sequence, labelled biochemically and viewed microscopically.

Fig. 4 shows the dependence of the redox potential on the hydrogen partial pressure at neutral pH. For reducing $SO_4^{2-}$ to $H_2S$, a redox potential of less than -200 mV is required (Tab. II); $CH_4$ oxidation to $CO_2$ works at a minimum redox potential of about -250 mV. These values narrow the field where microbial sulfate reduction can occur. The field is indicated by the two horizontal lines and arrows in Fig. 4. Due to the presence and anaerobic corrosion of steel / iron in a disposal, significant hydrogen partial pressures are formed. In the case of a disposal at a depth of several 100 m, the hydrogen partial pressure can surmount 1 bar. At these hydrogen pressures, only few microbial reactions are possible, such as the $CO_2/CH_2O$ and the $CO_2/CH_3OH$ at -430 mV and 390 mV, respectively.

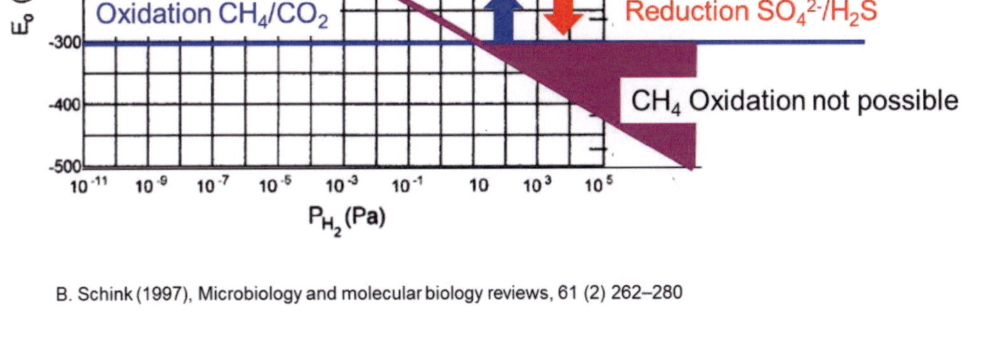


Fig. 4 Dependence of the redox potential on the prevailing hydrogen partial pressure at pH 7.0.

Reduction of ferric iron oxides, consumption of methane and formation of iron sulfides have been observed in methane-rich deep Baltic Sea and Black Sea sediments. Such processes are considered to limit the methane release from the sea sediment into the ocean water [17, 18].

Unfortunately, in these publications, the pH and redox potentials of the sediment pore waters are not reported.

## 3.4 pH range of microbial activity

The acidity or alkalinity of an environment greatly affects microbial growth (details are provided in various text books, e.g. [12] or [19]). Every microorganism has a pH range within which growth is possible, and growth shows a well-defined pH optimum. The pH range of growth is typically in a range of ± 1 to 1.5 around the optimum pH. Through their metabolic activity, microorganisms produce acid or alkaline metabolites which are released from the cell. These metabolites have the potential to change the pH of the medium. However, the internal pH of the cells needs to remain relatively close to neutral, even if the external pH is highly acidic or alkaline. Due to diffusion through the cell membrane, the cell requires a constant buffering of the internal pH. Most prokaryotes die, if the internal pH drops much below 5.0 to 5.5. Buffer compounds are phosphate buffers (neutral), borate buffers (alkaline), and citrate buffers (acidic).

Different strains of microorganisms are found in a wide pH range.

- Acidophiles (e.g. *Helicobacter pylori*) have an optimum pH range between 0 and 5.5. These species are found in sulfuric pools and geysers, areas polluted by acid mine drainage and even in the human stomach. For example, acidophiles oxidize reduced sulfur compounds to sulfuric acid and were found growing at pH = 0 in acid mine drainage in Iron Mountain in California.

- Neutrophiles (*E. coli*) have an optimum pH range of 5.5 - 8. The majority of these species are found in soil and water as well as in blood and tissue. When neutrophiles are exposed to a low pH environment, a mechanism is activated in order to reduce the effect of the decreasing internal pH by induction of glutamate decarboxylases, and other enzymes.

- Alkaliphiles (*Bacillus alcalophilus*) shows an optimum in pH range 8 - 14. Its internal pH is slightly acidic in comparison to the pH of the environment. The internal pH is maintained by specific regulators (homeostasis). This regulation is a highly energy-consuming physiological mechanism.

In Tab.III, pH optimum for growth of species of the physiological classes from acidophile to neutrophile and alkaliphile are listed.

Tab. III pH optimum for growth of some microorganisms [12].

| Physiological Class (optimal pH range) | pH optimum for growth | Example |
|---|---|---|
| Acidophile<br>pH < 5.7 | 1<br>3<br>5 | *Picrophilus oshimae* (archaea)<br>*Acidithiobacillus ferrooxidans*<br>*Rhodopila globiformis* |
| Neutrophile<br>$5.5 \leq pH \leq 8$ | 7 | *E. coli* |
| Alkaliphile<br>$pH \geq 8$ | 8<br>9<br>10 | *Chloroflexus aurantiacus*<br>*Bacillus firmus*<br>*Natronbacterium gregoryi* (archaea) |

## 3.5 Ionic strength and chaotropicity

The archaeal and bacterial window of life is restricted to a certain availability of water (in solution). The availability of water in a solution is expressed in terms of the dimensionless water activity, $a_w$. In a diluted solution $a_w = 1$ and in a saturated NaCl solution $a_w = 0.75$. Microorganisms are extremely dependent on exchange processes with their environment. This means that the macromolecules of the hydrophobic cell membranes need to function. If the hydrophobic state is changed, the characteristic steric structures of macromolecules change, leading to the denaturation of the respective macromolecule. The hydrophobic state of the macromolecules is controlled by hydrogen bonds. The regions of growth for different types of microorganisms is presented in Fig. 5. Optimum NaCl concentration for the marine microorganism (*Aliivibrio fischeri*) is in the range of about 3%. Extreme halophiles such as *Halobacterium salinarum* can exist in concentrated NaCl solutions between 15 und 30% (30% NaCl: ~5 mol/l).

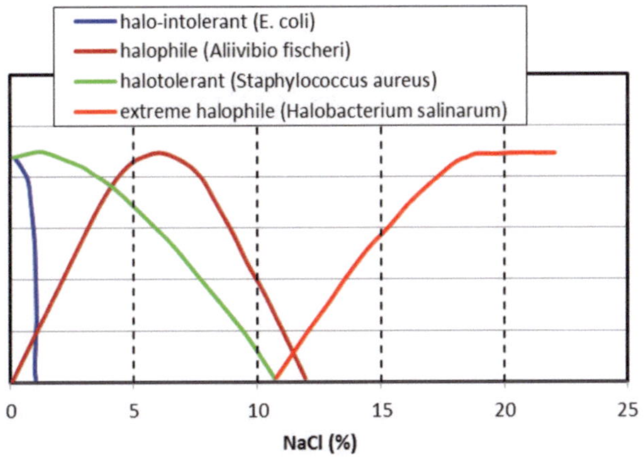


Fig. 5 Effect of NaCl concentration on the growth of microorganisms of differently pronounced salt tolerance or different need of NaCl [12].

A measure for the functioning of hydrogen bonds is the chaotropicity. Cations and anions disturb the hydrogen bonds of the macromolecules. The effectiveness of ions on the disturbance of the macromolecules in water is classified in the Hofmeister series.

- Anions: $SO_4^{2-} < HPO_4^{2-} < Cl^- < NO_3^- < Br^- < ClO_3^- < I^- < ClO_4^- < SCN^-$
- Cations: $NH_4^+ < K^+ < Na^+ < Li^+ < Mg^{2+} < Ca^{2+}$

In the Hofmeister series, the disturbance of the hydrogen bonds or chaotropic impacts increase from left to right. The ions at the left-hand side counteract the chaotropic impacts and are called kosmotropic. The kosmotropic ions intensify the hydrophobic effects of cell membranes by promoting the aggregation of proteins. Ions at the right-hand side are called chaotropic and decrease the hydrophobic interactions. If the hydrophobic state is changed, the characteristic steric structures of macromolecules change, leading to denaturation of the macromolecule. Chaotropic substances increase solubility of hydrophobic molecules, reduce enzymatic activities, disturb lipid double layers of the cell membranes and consequently cause stress for living cells.

Hallsworth et al. [20] showed "*that the chaotropicity of $MgCl_2$ at high concentrations not only denatures macromolecules, but also preserves the more stable ones: such indicator molecules, hitherto regarded as evidence of life, may thus be misleading signatures in chaotropic environments. The chaotropicity of $MgCl_2$ would appear to be a window-of-life-determining parameter, and the results obtained here suggest that the upper $MgCl_2$ concentration for life, in the absence of compensating solutes, is about 2.3 M (or above 2.5 M in the presence of a chaotrope-counteracting kosmotrope, such as NaCl).*" Other investigations of deep sea microorganism showed microbial activities up to solute concentrations of 3.2 mol/l [21, 22].

The chao-/kosmotropic activities were calculated according to differences in heat capacity of agar solutions, and expressed in kJ $kg^{-1}$ $mol^{-1}$. The resulting scale is numerically continuous and chaotropic compounds ranged from $MgSO_4$ (-64.5), NaCl (-11.0) to $MgCl_2$ (+54.0) and $CaCl_2$ (+92.2) [23].

# 4 Microbial effects in repositories

In the introduction chapter, four questions were asked. In Chapter 3, it was shown that microorganisms occur naturally in deep geological formations and that they could survive under conditions after repository closure. However, these conditions depend strongly on the disposal concept and need to be discussed separately for each disposal safety case. In Germany, four disposal concepts have been elaborated. These concepts include

- The Asse II salt mine near Wolfenbüttel is an approximately 100-year-old potash and salt mine. Between 1967 and 1978, 46 950 cubic meters of low- and intermediate-level radioactive waste in 125 787 drums were emplaced. Later, the mine was used to test the handling and storage of radioactive waste in a repository. The Asse mine faces two major problems: The penetration of saturated saline solutions (NaCl) and the stability of the mine openings, which is endangered.

- The Morsleben repository ERAM was a potash and rock salt mine. In 1971, the GDR[c] established a repository for low-level and intermediate-level radioactive waste in the mine. The Federal Republic of Germany continued to use this repository until 1998. Altogether 36 754 cubic meters of radioactive waste have been stored. The repository is under decommissioning. The objective is to safely seal off the radioactive waste from the biosphere.

- The "Mine for the Exploration of the Salt Dome Gorleben" has been investigated since 1979 for its suitability as a repository for highly radioactive waste. The two shafts Gorleben 1 and 2 reach a depth of 933 m and 840 m. The salt dome has a size of approximately 14 km length and 4 km width. Safety analyses (SA) were performed assuming different disposal concepts [24]. The first SA took place in the scope of a project including all components of the planned German disposal center at Gorleben, the Projekt Sicherheitsstudien Entsorgung (PSE) [25]. The disposal concept was based on a primary energy consumption of 60 000 TWh. This number corresponded to an operational period of some 160 years.

  In 1980 first calculations and sensitivity analyses were performed at KfK analyzing the radionuclide release from a flooded rocksalt disposal [26]. In this study, the convergence of the rocksalt was assumed to be the driving force for the release of contaminated brines into the aquifer system.

  A second SA considered only high-level waste disposal and was performed in a project funded by the European Commission (PAGIS) [27]. A third SA took into account a

[c] GDR: German Democratic Republic

significantly smaller disposal capacity as well as the disposal of vitrified HLW and spent nuclear fuel elements, Systemanalyse Mischkonzepte (SAM) [28].

- In contrast to the previous disposals, the Konrad mine is an abandoned iron ore mine located in the area of the city of Salzgitter. It is currently being converted to a repository for radioactive waste with negligible heat generation. About 90 per cent of the radioactive waste accruing in Germany is in this category. The mine is in a depth of 800 to 1 300 meters in an iron ore deposit. Compared with other iron ore mines, Konrad is exceptionally dry. The covering layer of clay rocks, which is up to 400 m thick, seals the mine against the groundwater.

In this chapter, an answer will be given on the question of effects by the environmental conditions in a deep nuclear waste repository on resident or introduced microorganisms and their performance. The information was based on literature mainly, concerning the degradation of the waste forms, the corrosion of metallic materials and the alteration of the Eh and pH as well as the generation of complexing agents by degradation of organic material. Most important for the SA models was the expected gas production ($CO_2$, $H_2$, $CH_4$) leading to an increase of the pressure and consequently a promotion of the driving forces for radionuclide releases.

## 4.1 Degradation processes of waste components

### 4.1.1 Microbial Degradation of Paper, Plastics and Textiles

In the Waste Isolation Pilot Plant (WIPP, USA) transuranic waste bearing organic matter to some extent is emplaced. There is a strong interest in estimating the production of total inorganic carbon, TIC, that could be produced by degradation of organic waste. Francis and Gillow [29-31] examined the biodegradation of mixed cellulosic and other organic materials such as polyethylene (PE), polyvinylchloride (PVC), hypalon, and neoprene in G-Seep brine (natural brine from the WIPP site) and determined the total gas production and the gases $CO_2$, $N_2O$, $H_2$, $O_2$, $CH_4$ and $H_2S$. The effects of variables such as

- initial atmosphere (air or nitrogen),
- water content (humid (~70% relative humidity) and brine inundated), and
- nutrient amendments (nitrogen, phosphate, yeast extract, and excess nitrate)

were investigated. The gas production was determined by pressure measurement, $CO_2$ and $CH_4$ were analyzed by gas chromatography. The dissolved cellulose degradation products were analyzed by high-performance liquid chromatography and the microbial populations in the samples were determined by direct microscopy and molecular analysis. The results reported by Gillow [31] were considered in German disposal safety analyses (Tab. IV).

Tab. IV shows clearly, that in the presence of sufficient nutrients and especially nitrate, the metabolic rates increase significantly. Francis and Gillow [31] analyzed the DNA of the microorganisms from cellulose samples in the brines. DNA showed a diverse assemblage of bacteria and archaea in the presence and without additional nutrients. Cellulose degradation products found in the solution include fumaric, lactic, oxalic, oxalacetic, propionic, and succinic acids indicating fermentative microbial activity. The presence of bentonite enhanced the total gas production, concentration of gaseous and aqueous metabolites. Microbial gas production was not observed from plastic and rubber materials after about 7 years' incubation.

Tab. IV Gas production from microbial cellulose degradation over a period of 10.8 years [31].

| | | Nutrient | $NO_3$ | Humid conditions | Inundated |
|---|---|---|---|---|---|
| **aerobic conditions** | Gas | - | - | | 0.84 ± 0.10 mL/g |
| | | + | - | | 1.71 ± 1.03 mL/g |
| | | + | + | | 12.2 ± 0.0 mL/g |
| | $CO_2$ | - | - | 6.1 ± 2.4 μmol/g | 16.3 ± 1.3 μmol/g |
| | | + | - | 0.5 ± 0.3 μmol/g | 41.4 ± 7.8 μmol/g |
| | | + | + | | 186 μmol/g |
| | $CH_4$ | - | - | | 1.34 ± 0.03 nmol/g |
| | | + | - | | 0.84 ± 0.05 nmol/g |
| | | + | + | | 1.27 ± 0.37 nmol/g |
| **anaerobic conditions** | Gas | - | - | | 2.48 ± 0.31 mL/g |
| | | + | - | | 4.12 ± 0.76 mL/g |
| | | + | + | | 18.1 ± 0.38 mL/g |
| | $CO_2$ | - | - | 115 ± 20 μmol/g | 27.4 ± 5.8 μmol/g |
| | | + | - | 22 ± 3.3 μmol/g | 66.9 ± 1.1 μmol/g |
| | | + | + | | 251 ± 5 μmol/g |
| | $CH_4$ | - | - | | 5.89 ±1.30 nmol/g |
| | | + | - | | 2.74 ± 0.90 nmol/g |
| | | + | + | | 2.57 ± 0.79 nmol/g |
| | Number | - | - | $(5.1 \pm 3.4) \times 10^5$ | $(1.59 \pm 0.15) \times 10^7$ |
| | Of | + | - | | $(1.62 \pm 0.07) \times 10^8$ |
| | Cells | + | + | | $(2.24 \pm 0.24) \times 10^8$ |

Concerning the number of cells, it needs to be mentioned that the cell counts were made at only one time point using a method that is not currently recommended for archaea. The DNA analysis was performed by "Denaturing Gradient Gel Electrophoresis" (DGGE) which separates genes of the same size based on their different denaturing ability. DGGE analysis was also applied only at a single time point with only one set of incubations. Furthermore, the DGGE was performed late during the incubation period at a time when much of the bacterial diversity would be missed. The DGGE results allow the interpretation that fermentation and nitrate reduction were the chief modes of microbial activity in these experiments. Unfortunately, the DNA sequences were not submitted to the international databases.

The gas production was recalculated in gas production rates. The total $CO_2$ production rates reported by Gillow [31] are given in Tab. V for the anaerobic experiments

Tab. V Rate of carbon dioxide production by anaerobic inundated samples

| Treatment | Overall rate of $CO_2$ Production µ-moles $g^{-1}$ cellulose $day^{-1}$ |
|---|---|
| Unamended/Uninoculated | $1.58x10^{-4}$ |
| Unamended/Inoculated | $6.44x10^{-3}$ |
| Amended/Inoculated | $1.39x10^{-2}$ |
| Inoculated + Excess Nitrate | $5.57x10^{-2}$ |

These numbers in Tab. V took into account the whole $CO_2$ formation from the beginning of the experiments ([31], page 21). Fig. 6 shows clearly that the $CO_2$ production is very high in the case of the samples with nutrients and excess nitrate in the beginning. Recalculating the $CO_2$ production rates for the long-term part of the curves (from 730 days to termination of the experiments) indicates significantly lower production rates ($6.4x10^{-3}$ µmol $g^{-1}$, for samples without nutrients and $1.4x10^{-2}$ µmol $g^{-1}$ for samples with excess nitrate).

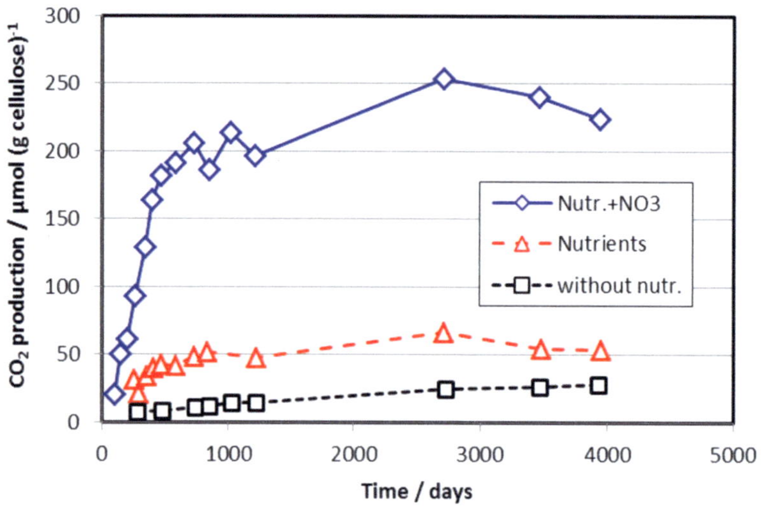


Fig. 6 Carbon dioxide produced in anaerobic samples inundated with brine: unamended and inoculated (□); amended and inoculated (Δ); amended, inoculated, plus excess nitrate (◇).

The data shown in Fig. 6 need to be balanced with respect to the available nutrients and nitrate: The experiments have been performed using 5 g cellulose per 100 ml solution, the nutrient addition amounted to 0.25 g/l and the excess nitrate to 0.5 g $KNO_3$ per 100 ml (50 mmol/l or 1 mmol $NO_3^-$ per g cellulose) [29]. Assuming a $C_{12}H_{20}O_{10}$ as sum formula of cellulose, the carbon content amounts to about 37 mmol carbon per gram.

In the case of the experiment indicated as amended, inoculated, plus excess nitrate (blue diamonds in Fig. 6), after about two years, the $CO_2$ production reached 0.2 mmol per g cellulose. This number indicates that neither the cellulose nor the excess nitrate was consumed and for this reason one could expect a continuation of the initial production rate. However, the production rate decreased.

Under conditions of the experiments, all cellulose will be degraded after about 10000 to 20000 years. These numbers provided the basis for the Asse II safety analyses. In the Asse II mine, about 8% of the waste canisters contain cellulose, paper, wood and cotton slightly compacted without further solidification [9]. The mass of the cellulose containing materials was estimated at 1200 Mg.

### 4.1.2 Microbial Degradation of Bitumen

In 1978, C. E. ZoBell and M. A. Molecke published a survey on microbial degradation processes especially on the degradation of bitumen in the context of WIPP [32]. The survey described the chemical and physical properties of bitumen and how these properties affect the degradation processes. Many of the bitumen components, such as asphaltenes, resins and paraffinic, napthenic and aromatic hydrocarbons have been shown to be more resistant against strong chemical reactants such as strong acids or alkaline solutions in comparison to microbial enzyme systems. Many microorganisms have the ability to produce alcohols, esters, ketones, and other fermentation products in which asphalts are soluble, albeit only slightly. Such solubilization may be of negligible significance during a period of several years, but could result in the perforation or partial matrix solubilization of the bitumen within decades or centuries, in an environment where water and organic matter (celluloses, proteins, lipids, etc.) are present along with elements essential for microbial growth, some solubilization of bitumen by fermentation products can be expected. ZoBell & Molecke [32] presented various methods for detection of bio-degradation of bitumen. Following conclusions were drawn:

Only a few species attack bitumen, such species appear to be widely distributed in soil, especially in contact with crude oil or bitumen. Bitumen-oxidizing bacteria have *been* incriminated in the deterioration of asphalt-paved streets and asphalt coating on buried pipes or other steel structures. This suggests that under certain conditions of storage where microorganisms, minerals essential for their growth, and a little moisture are present, bituminous coatings might be breached by bacteria within a few decades.

Even under the most favorable conditions for the microbial degradation, the rates are generally very slow. It usually requires several days to a few weeks to demonstrate the microbial degradation of bitumen. The rate is influenced primarily by the chemical composition of the bitumen. The rate of the degradation appears to be directly proportional to the surface area exposed to oxygenated water. Certain oil-oxidizing bacteria may contain enough water to enable the oxidizing oil, resulting in the formation of enough water to keep them growing:

$$C_{15}H_{32} + 23\ O_2 = 15\ CO_2 + 16\ H_2O \qquad (2)$$

By increasing the contact between enzymes and substrate, certain detergents tend to increase the rate of oil or bitumen oxidation. Many species of bacteria produce surfactants. The microbial production, properties, and functions of surfactants, with particular reference to petroleum, have been reviewed in a series of publications. The biodegradation of asphaltenes and other hydrocarbons by *a Ps*eudomonas species was found to be promoted by bio-emulsifiers or surfactants. *Corynebacterium lepus* and *Pseudomonas asphaltenicus* are species of bacteria isolated from asphalt deposits in Ontario, Canada. Both species attack higher hydrocarbons and produce surface-active substances that substantially reduce the interfacial tension of saline solution layered with kerosene. Although the optimal growth was on n-decane, *P. asphaltenlcus* was shown to utilize a wide range of hydrocarbons, including asphaltenes which make up an appreciable part of most bitumens.

Hyperbaric oxygen, $CO_2$, and certain other gases are bacteriostatic or bactericidal at pressures of only 0.5 to 1 MPa. Within the range of 0.1 and 30 mg/liter, the concentration of dissolved oxygen has no discernible effect on the rate of hydrocarbon oxidation by microorganisms. However, the rate of such oxidation is appreciably slower in the absence of free oxygen than when some oxygen is present. The microbial oxidation of hydrocarbons under strictly anaerobic conditions has been investigated by several authors. Although radiolytical degradation of water can produce hydrogen and oxygen. Studies on the radiolytical degradation of cellulosic transuranic contaminated waste materials have indicated that oxygen (in air) originally contained in the waste packages was depleted during the formation of CO and $CO_2$.

In the case of anaerobic degradation of bitumen, the rate would be much slower than aerobic degradation. Microorganisms grow in natural conditions and in the laboratory at redox potentials ranging from Eh -450 to +850 mv. The activity of iron-oxidizing bacteria is normally found in an environment with an Eh range limited to $+60\ V \leq Eh \leq 850\ mV$ at one extreme; sulfate-reducing bacteria are active in the range of $-450 \leq Eh \leq +50\ mV$.

In order to evaluate the microbial degradation of bitumen, Ait-Langomazino published investigations on following topics [33]:

a) Quantification of the rate of biodegradation by determination of the $CO_2$ production.
b) Showing the possible alterations of bituminous material through chemical transformations such as the oxidation processes.
c) Comparing the activities of microorganisms from the metabolic specificity.
d) Determination whether the chemical nature of bitumen can affect the degree of microbial utilization by comparison of two types of bitumen:
   blown Mexphalte R 90/40 and direct distillation Mexphalte 80/100.

Degradation was tested with various pure strains: *Pseudomonas aeruginosa, Pseudomonas putida, Arthrobacter sp., Saccharomycopsis lipolytica* (DSM Nos 1253, 50202, 312 and 1345, respectively); *Torulopsis holmii* was isolated from a fuel tank and a natural mixed culture enriched from a bituminous site which included a strain of *Pseudomonas aeruginosa.*

The biodegradation potentials of four pure cultures, bacteria and yeasts, and three natural or reconstituted mixed cultures on a Mexphalte R 90/40 in aerobic batch conditions were found to be very similar. From $CO_2$ generation, maximal degradation of bitumen R 90/40 was estimated to be 9.0 wt.% $3.2\times10^{-3}$g cm $^{2}$ and 3.1 - $9\times10^{-3}$ cm for *Saccharomycopsis lipolytica.* It was assumed that the cessation of $CO_2$ production was real, that bitumen degradation occurred through homogeneous surface attack and that the surface area of the bitumen was homogeneous. Metabolic activity in batch culture was stopped probably due to inhibiting conditions present, e.g. accumulation of metabolic byproducts.

The Mexphalte R 90/40 degradation rate by *S. lipolytica* was determined by the $CO_2$ generation. It was closely linked with a biofilm formation at the surface of the bitumen sample. Of the various chemical fractions of Mexphalte R 90/40 the saturated and the aromatic fractions were lost with about a 40% and 25% reduction in these fractions respectively. A complete disappearance of n-paraffins, an increase of oxygen content in the resins and the asphaltenes characterize bitumen oxidation by *Saccharomycopsis lipolytica.*

Growth on individual bitumen fractions confirms the preferential utilization of the n-saturates fractions. Resins and asphaltenes are resistant to *Saccharomycopsis lipolytica* attack and their presence inhibits the biodegradation of n-saturates. The chemical composition of bitumen has a marked effect on its biodegradability. Direct distillation bitumen Mexphalte 80/100 retained for the coating of LLW wastes shows a greater resistance to degradation by *Saccharomycopsis lipolytica* than Mexphalte 90/40 probably due to its low content of saturates. The composition and content of other bitumen components may also influence the biodegradability. It has been shown that bituminous materials are subject to microbial attack and deterioration in many of their technical applications. Therefore the use of bitumen materials for the coating of low and intermediate level wastes requires a deeper knowledge of the effects that the environmental conditions of a repository can have on its biodegradation [33].

In the Technical Report NTB 89-14 released by Nagra [34], Wolf et al. concluded that the presence of microorganisms in nuclear waste repositories cannot be ruled out. The potential of different microbial consortia to degrade bitumen has been examined. Experiments have been carried out under optimal culture conditions using bitumen with a highly-increased surface area. The investigations showed clearly that bitumen-degrading organisms are ubiquitous. In general, the organisms formed biofilms on the accessible substrate surface area. The long-term bitumen degradation rates were found to be essentially independent of the culture initially inoculated. Investigations were carried out both under aerobic and anaerobic conditions using a variety of cultures. Under aerobic culture conditions a bitumen degradation rate of 20 - 50 g bitumen $m^{-2}$ $yr.^{-1}$ leading to a $CO_2$ production of 15 - 40 l was observed. Anoxic conditions yielded a 100 times smaller degradation rate of 0.2 - 0.6 g bitumen $m^{-2}$ $yr^{-1}$ (STP) and a $CO_2$ production of 0.15 - 0.45 l (STP) [35].

Based on linear extrapolation the experimentally determined degradation rates would lead to a 25 - 70% deterioration of the bitumen matrix under aerobic and 0.3 - 0.8% under anoxic conditions within 1000 years [34]. These results are considered in the safety analysis for the

German Low-Level Radioactive Waste Disposal Morsleben (ERAM): In the mine, bitumen was used as a sealant in underground constructions [36]. The modeling of microbial degradation processes relied on Nagra Report NTB 89-14. According to reference [37] the microbial degradation front penetrates into the sealant construction some 6 µm $yr^{-1}$ under aerobic conditions.

Another study [38] investigated *Pseudomonas aeruginosa* and *Saccharomycopsis lipolytica* under aerobic conditions. The biodegradation of a bitumen sample was expressed as the amount of $CO_2$ produced during a twelve-month period with periodic oxygen and medium renewal. A linear production of $CO_2$ was observed in the case of the *P. aeruginosa* strain. The limitation of the amount of $CO_2$ produced by *S. lipolytica* could be explained by a limited rate of biodegradable hydrocarbons (probably the saturated fraction). This limitation was not observed with the *P. aeruginosa* strain and was explained by the potential of this bacterium to attack different hydrocarbons, for example the aromatic fraction.

### 4.1.3 Microbial Degradation of Compacted Mixed Wastes

Funded by the nuclear industry (Nukem GmbH, Hanau) in 1986, the Technical University Darmstadt, Institute for Microbiology elaborated a study on the probability of microbial activities in compacted mixed wastes under conditions relevant for the German LLW disposal Konrad [39]. These wastes should be compacted under a pressure above 30 MPa. Depending on the nature of the material, a compaction by a factor of 5.8 (average) was achieved [40]. Boundary conditions for the study were the available substrates, such as paper, wood, plastics, textiles, and the prevailing ambient conditions of 50°C temperature, 5% remaining humidity in the waste products and about 80% relative air moisture. Model experiments under simulated Konrad conditions covered following questions:

- Which kind of microorganisms (bacteria and fungi) survives under the ambient conditions.
- Is a microbial evolution possible under these conditions and can be detected by analyzing cell proliferations and/or metabolites.

In a first experiment, the survival of microorganisms on membrane filters was investigated. The membranes were inoculated with the microorganisms listed in Tab. VII and were stored under atmospheric conditions of 30, 80 and 95% relative humidity in closed preserving bottles. Nutrients for microorganisms were glucose, peptone, meat extract and yeast extract.

The pellets were prepared as follows: About 40 g of mixed simulated waste materials were compacted using pressures up to 40 MPa forming pellets of 5 cm diameter and 1.5 cm thickness. The pellets were seeded with the following bacteria and fungi (including yeasts). The experiments were performed using 4 pellets at 28°C and 50°C and for each relative humidity (33, 80 and 96%) up to 6 months. At several time intervals samples were prepared and cultivated on substrates in order to identify the colony forming microorganisms. Additionally, microflora from digested sludge was investigated at 32, 80 and 97% relative humidity at 37°C for 6 weeks. The humidity was maintained by the water saturation vapor of concentrated salt solutions in the lower part of the preserving bottles separated from the pellets (Tab. VI).

Tab. VI Relative humidity over saturated salt solution at 28°C and 50°C [41].

| Salt | 28°C | 50°C |
|---|---|---|
| $MgCl_2 \times 6H_2O$ | 33.0 | 31.4 |
| $(NH_4)_2SO_4$ | 80.8 | 78.1 |
| $K_2SO_4$ | 97.2 | 96.8 |

Tab. VII Microorganisms investigated with respect to their survival in compacted mixed wastes under Konrad conditions.

| Species | Gram Stain Reaction[d] | Cell form |
|---|---|---|
| *Escherichia coli* DSM 682 (= ATCC 10 536) | gram-negative | vegetative cells |
| *Pseudomonas fluorescens* | gram-negative | vegetative cells |
| *Azotobacter sp.* DSM 281 | gram-negative | vegetative cells + cysteine |
| *Staphylococcus aureus* DSM 346 (= ATCC 6538 P) | gram-positive | vegetative cells |
| *Arthrobacter sp.* | gram-positive | vegetative cells |
| *Mycobacterium sp.* | gram-positive | vegetative cells |
| *Streptomyces griseus* DSM 40 236 (= ATCC 23 345) | gram-positive | aerobic mycelia spores (conidia) |
| *Thermoactinomyces sp.* DSM 22 (= ATCC 12 980) | gram-positive | endospores |
| *Bacillus cereus* DSM 345 (= ATCC 11 778) | gram-positive | endospores |
| *Clostridium sporogenes* DSM 633 | gram-positive | endospores |
| *Alternaria* | filamentous fungus | conidia |
| *Cladosporium* | filamentous fungus | conidia |
| *Penicillium* | filamentousfungus | conidia |
| *Aspergillus niger* | filamentousfungus | conidia |
| *Mucor* | filamentousfungus | conidia |
| *Rhizopus* | filamentous fungus | conidia |
| *Rhodotorula a* | yeast | vegetative cells |

DSM: "Deutsche Sammlung von Mikroorganismen"

A first test was performed on membranes under the ambient conditions.

[d] The Gram Color is a method developed by Hans Christian Gram (1853–1938) which allows the dyeing of bacteria for microscopic investigations. Bacteria can be distinguished by their color in two main groups: gram-positive and gram-negative. These groups have different cell membrane properties.

- At 50°C none of the tested microorganisms survived longer than one month independent of the relative humidity. However, a thin crystalline salt layer was found to cover the membrane filters after termination of the experiments. These salt precipates might have caused stress to the microorganisms.
- The same result was obtained for *Escherichia coli*, *Azotobacter*, and *Staphylococcus aureus* at 28°C.
- Resistant against the exposure conditions was the aerobic spore former *Bac. cereus*. At 28°C it survived 5 months (the only species of the selected microorganisms) at 30% relative humidity. Another species *Bac. stearothermophilus* which is also an aerobic spore former showed the same sensitivity as the species *Escherichia coli*, *Azotobacter*, and *Staphylococcus aureus*. Also, the anaerobic spore former *Clostridium sporogenes* was only marginally more resistant.
- The fungi, both filamentous and the yeast *Rhodotorula,* showed relatively uniform behavior. At 28°C all species survived at 96% relative humidity for 5 months. At 33% relative humidity, the species died off after 3 months, and at 80% relative humidity only *Aspergillus niqer* survived for 5 months.
- The digested sludge didn't show a mass loss at 97% relative humidity, at lower humidity, the mass of the sample was reduced by ~10%. In a nutrient solution, all samples recovered within 12 hours showing digestion and gas formation about 10% $CO_2$ and 90% $CH_4$. A smell of butyric acid and $H_2S$ could be noticed.

A second experiment was performed with compacted granulated particles (pellets) consisting of paper, wood, plastics, textiles. The pellets were doped with microorganisms taken from different sludges of a sewage purification plant and sorted under the ambient conditions as described above for 6 months. Also these experiments were performed in bottles sealed with rubber septa and crimped metal rings.

- After 6 months, the 4 pellets were resuspended in sterile tap water, and viable cells in the aerobic and anaerobic supernatants were determined by plate counts. Then the vials were closed and stored at 37°C. The gas production in the vials was measured.
- The viable plate counts showed that the ambient conditions didn't change the microorganism population in the pellet. The authors assumed that the mixture of the sludge could protect the bacteria cells.
- After contacting the pellet with water, the water in the vials turned muddy and a digestion process took place which was not the case for un-doped pellets. Gas bubbles occurred within one week and a smell of butyric acid and $H_2S$ was noticed. The simulated waste pellets acted as substrate.
- It could also be shown that *Escherichia coli* and *S. aureus* were sensitive to the compaction process of the pellet.

In total, the microorganisms taken from the sludge were found to be more resistant against the ambient conditions of an interim storage or deep disposal conditions.

### 4.1.4 Microbial Attack of Cement

Three groups of organisms have been identified that create conditions that can compromise concrete integrity. They include sulfur oxidizing bacteria, nitrifying bacteria, and some heterotrophic organic acid producing bacteria. Species of each of these groups of organisms have been isolated from degrading concrete [42]:

- Sulfur oxidizing bacteria (genus *Thiobacillus*) are the microorganisms most often associated with the biological degradation of concrete structures. At pH ~ 9, *Thiobacillus thioparus* actively begins to oxidize available forms of reduced sulfur. Products from this process contribute to a continued decrease in surface pH.
- Nitrifying bacteria are known to promote concrete degradation. These bacteria (*Nitrosomonas* and *Nitrobacter*), obtain energy through the oxidation of inorganic nitrogen compounds. Neutral to alkaline soils have the largest populations of nitrifying bacteria. Both the ammonium oxidizing *Nitrosomonas sp.* and the nitrite oxidizer *Nitrobacter sp.* have been isolated from degraded concrete.
- Heterotrophic microorganisms capable of producing organic acids through the assimilation of organic carbon compounds can be found everywhere. Organic acids such as lactic, citric, gluconic, malic and many others are byproducts of the metabolism of these microbes. Because of their diversity, heterotrophs can exist under a much wider range of environmental conditions than either the sulfur oxidizers or nitrifying bacteria.

The biodegradation of cement is due to the products of microbial metabolism, such as organic or mineral acids, and thus is an indirect effect of microbial growth. Microorganisms can produce organic acids resulting from cellulose biodegradation; cellulose is representative of an organic solid waste. In the same way, microorganisms can produce mineral acid ($H_2SO_4$) from sulfur, as a source of energy; sulfur can be present in the environment of nuclear waste, or in the waste itself [38].

Experiments were performed with a medium consisting of approximatively 15 wt.% of portlandite $Ca(OH)_2$, immersed in a mineral medium. This medium was inoculated and stored for a period of 12 months. The microbial inocula consisted of

- A mixed culture of fungi, (*Aspergillus niger*, *Trichoderma viride*), heterotrophic microorganisms producing organic acids isolated from a soil sample. The carbon and energy source was glucose, which is an intermediate product of cellulose degradation. The initial pH was about 9.
- A mixed culture of *Thiobacillus*, autotrophic microorganisms producing $H_2SO_4$ from sulfur as an energy source. The initial pH was 7.

An increase of pH to pH = 12 due to leaching of $OH^-$ from the concrete was observed. This increase of pH inhibited, for an 80 day period the growth of *Thiobacillus*, which was shown by the lack of $H_2SO_4$ production. In a second experiment, with fungi, an immediate organic acid production was observed, indicating a rapid adaptation of these microorganisms to basic pH. In both cases, the acid production was correlated to the amount of leached calcium.

Organic acids were quantitatively and qualitatively determined during growth of fungi over a two-year period. Even with high pH conditions, pH ~ 11, growth of microorganisms occurred.

## 4.2 Microbial activities in high ionic strength systems

In 2016, a comprehensive summary was provided by Swanson et al. elaborating the microbiology of subsurface, salt-based nuclear waste repositories [43]. This report reviews a series of publications[e] dealing with the microbial activities at high ionic strength solutions. An answer was given to the question: "Is there a common water-activity limit for the three domains of life[f]?" [44]. The authors of this study conclude that the water availability in terms of water activity determines both the vitality and functionality of living systems. The majority of microbes cannot multiply below 0.900 $a_w$ and for the most extremophilic species, cell division has only been observed down to ~0.61 $a_w$. The established water-activity window for cell division of archaeal and bacterial life (1 - 0.755) is narrower than that of some xerophilic fungi that are even able to grow and/or germinate in the range 0.755 - 0.605 $a_w$. Halophilic archaea and bacteria are capable of cell division down to the 0.605 water-activity limit. The reason for the archaeal and bacterial life window at $a_w$ above 0.605 consists in the behavior of the macromolecules of the hydrophobic cell membranes (see Chapter 3.5).

With reference of Hallsworth [20] and Cray [23], Swanson et al. [43] tried to predict microbial effects on the WIPP repository by analyzing the microbial ecology, bioenergetics, and conditions. She made use of Hallsworth's graph showing the window-of-life for microorganisms in the water activity vs. concentration field (see Fig. 7).

---

e References are given in the respective literature.

f Bacteria, Archaea and Eukarya (see Appendix Tab. XIV Characteristics of the three domains of life [12, 84] .)

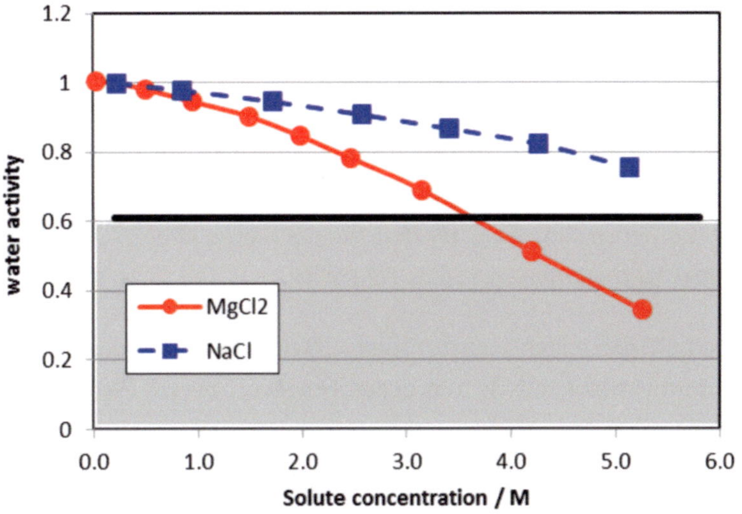


Fig. 7 Water activity reduction over a range of $MgCl_2$ and NaCl concentrations at 25°C. The white region indicates the water activity range for microbial function (0.61 – 1).

For a Safety Case, the findings obtained from pure solutions need to be transferred to disposal systems. In Germany, most of the I/LLW wastes are cemented. The interactions between cemented waste forms and NaCl- and $MgCl_2$-rich solutions have been investigated for periods of more than 30 years. Details of the experiments, water to cement ratios of full-scale cement blocks, brines compositions and mass to volume ratios are described in [45].

Tab. VIII Characteristic data and final solution compositions of selected full-scale cement samples corroded in NaCl- and $MgCl_2$-rich solutions [45].

| | **Initial NaCl solution** | **#35** | **Initial $MgCl_2$ solution** | **#25** | **#29** | **#33** | **#36** |
|---|---|---|---|---|---|---|---|
| **w/c** | - | 0.5 | - | 0.30 | 0.43 | 0.50 | 0.50 |
| **m/v in kg/l** | - | 2.5 | - | 2.5 | 1.8 | 2.2 | 2.1 |
| **Duration** | - | 26 yrs. | - | 29 yrs. | 29 yrs. | 17 yrs. | 24 yrs. |
| | | | | **Mol $l^{-1}$** | | | |
| **$MgCl_2$** | 0.04 | 0.00 | 3.79 | 4.50 | 0.14 | 0.04 | 0.00 |
| **$CaCl_2$** | 0.03 | | 0.00 | 0.00 | 2.71 | 2.90 | 2.14 |
| **NaCl** | 6.10 | 4.73 | 0.38 | 0.42 | 1.75 | 0.60 | 1.09 |
| **$NaNO_3$** | - | 1.53 | 0.00 | 0.97 | 1.73 | 2.14 | 2.02 |
| **KCl** | 0.04 | 0.17 | 0.71 | 0.36 | 0.71 | 0.96 | 0.72 |
| **$MgSO_4$** | 0.001 | 0.01 | 0.29 | 0.34 | 0.00 | 0.01 | 0.00 |
| | | | | | | | |
| **$a_w$** | 0.77 | 0.77 | 0.55 | 0.47 | 0.59 | 0.62 | 0.70 |
| **$-\log m_{H+}$** | 8.2 | 13.5 | 8.8 | 9.7 | 11.9 | 12.8 | 11.9 |

Application of the chaotropic compounds reported by Hallsworth [23] a correlation between the chaotropicity and the water activity for the cement attacking solutions given in Tab. VIII was obtained. The correlation is shown in Fig. 8.

The chaotropicity shown in Fig. 8 was calculated by adding the product of the chaotropicity data [23] times the ion concentrations given in Tab. VI. The chaotropicity of $NaNO_3$ was assumed to be zero. The figure shows clearly that the calculated chaotropicity are close to the expected values defined by the initial corroding solutions.

- Solution of sample #35 which was corroded in NaCl solution revealed a chaotropicity in the field where microbial activity may occur. However, the pH (-log $m_{H+}$) of this sample was outside of the window of life.
- Sample #36 was corroded in the $MgCl_2$-rich solution. Due to the calcium-magnesium exchange reaction, the solution changed from magnesium dominated to calcium dominated regime. Furthermore $NaNO_3$ was released from the simulated cemented waste sample to the solution. In total, the water activity decreased. Consequently, microbial processes cannot be excluded under such conditions.
- The solutions of samples #29 and #25 show water activities outside of the window-of-life ($a_w < 0.61$). Water activity of solution from sample #33 is $a_w = 0.62$. However, in this case the chaotropicity is significantly above the value of a pure $MgCl_2$ solution due to the high calcium concentration. The calculated chaotropicity of solution #33 would correspond to a water activity of 0.45 in a pure $MgCl_2$ solution. For this reason, the occurrence of microbial activity seems to be extremely implausible.

Additional to the constraints of microbial life by the chaotropicity, the pH plays a significant role. Only very few species survive under pH conditions in the range of 12.

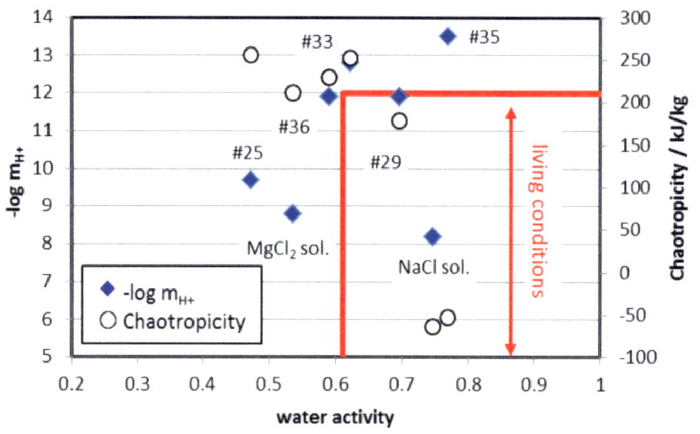


Fig. 8 pH range and calculated chaotropicity (data from [23]) versus the water activity of the NaCl-rich (squares) and initially $MgCl_2$-rich solutions (bullets). Red line: window-of-life for $a_w > 0.61$ and pH < 12.

# 5 Microbial effects on barriers and migration

Another question mentioned in the introduction concerned the effects of microorganisms on a repository in terms of structural integrity of the engineered barriers, and on radionuclide release and subsequent migration. In order to answer this question, the microbiological characterization of the cap rocks above the Gorleben salt dome was analyzed and the effects on the sorption behavior of the redox sensitive radionuclides technetium and selenium was investigated. Concerning the structural integrity of the engineered and natural barriers, microbial sulfate reduction was investigated theoretically. In the case of microbial sulfate reduction, a reduction of the specific volumes by 20 vol.% occurs when the mineral anhydrite ($CaSO_4$) is transformed to calcite ($CaCO_3$). This process was extensively analyzed in the scope of the Preliminary Safety Analysis Gorleben (vSG).

## 5.1 Cap rock above the Gorleben salt dome

For the safety assessment of the radioactive waste disposal planned at the Gorleben salt dome, the barrier functions of the relevant geological formations were analyzed since 1981. Specific interest was focused on the overlying strata of the salt dome. The investigations included sorption and desorption studies for various radionuclides, complexation of radionuclides with groundwater ligands, the behavior of colloids and the influence of microbial activity on the potential migration of radionuclides. Preliminary results have been reported and were published in a specialist meeting "Radionuclide migration in the cap rock above the Gorleben repository" [46].

The contribution on Microbiologic Investigations of Soil Samples of Overlaying Strata of the Gorleben Salt Dome" were elaborated by H. J. Kutzner, Technical University Darmstadt [47, 48]. The results were summarized as follows:

1. During the drilling of shaft II into the Gorleben salt dome, soil samples were collected at depths between 115 to 170 m. The samples were characterized microbiologically as well as biochemically. For shaft drilling, the soil had been frozen for several years and the samples were kept at -20°C until they were processed.
2. A specially designed device was constructed to remove sterile samples from the frozen soil. Although all measures have been carried out under sterile cautery, a small contamination of the material with "environmental germs" during sampling in the shaft, packaging in plastic bags and subsequent handling could not be completely ruled out.

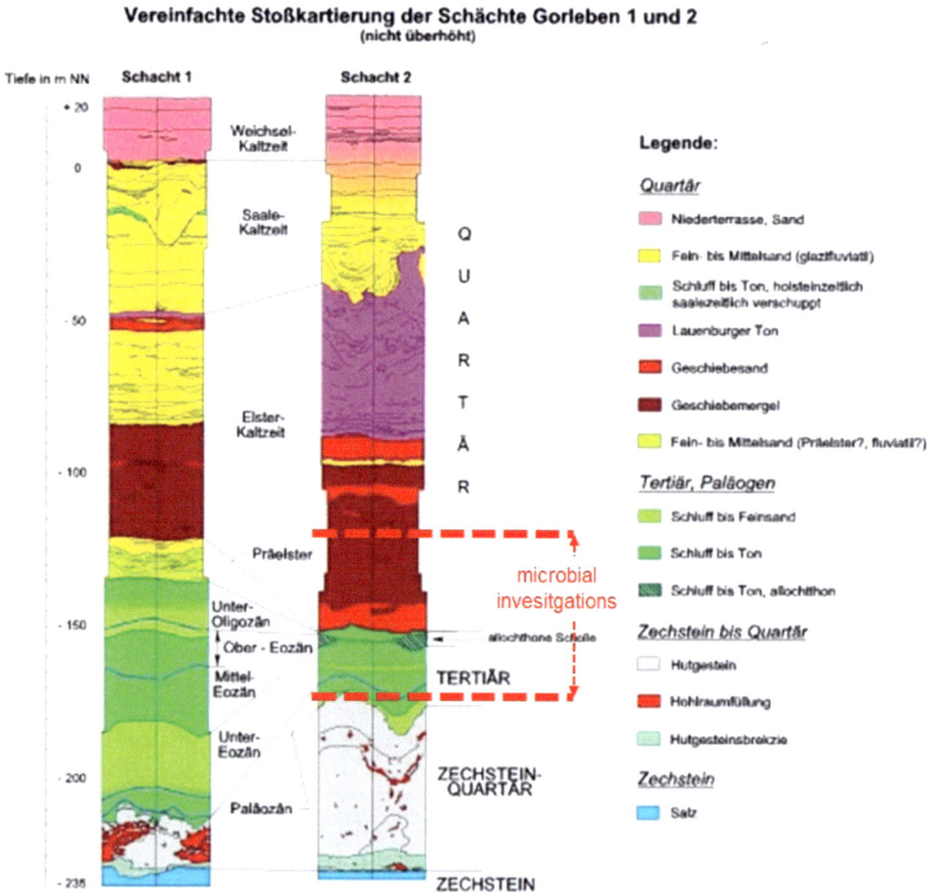


Fig. 9 Geological horizons of the Gorleben cap rocks. Red lines indicate the sampling locations for the microbial investigations at shaft II. (http://www.deutsche-rohstoffagentur.de/DE/Themen/Wasser/Projektbeitraege_NIL/Endlagerung_Gorleben/geol_schaechte_fb.html, 2016).

3. The microbiological characterization showed that the soil samples had an extremely low microbial content of heterotrophic bacteria: No bacterial cells could be detected with the help of microscopic methods (light & SEM). During plating on various nutrient media, the undiluted samples produced only a few colonies per agar plate. (For comparison: a fertile garden soil contains $10^8$ to $10^9$ bacteria per g). Therefore only 26 bacteria could be isolated, most of them were identified with known bacterial species. Following bacteria were identified: *Kytococcus sedentarius, Micrococcus sp., Bacillus cereus, Bacillus circulans, Brevibacterium linens, Aeromonas hydrophila, Suttonella indologenes, Brevundimonas vesicularis, Brevundimonas diminuta, Pseudomonas alcaligenes, Delftia acidovorans, Acetobacter pasteurianus*, and *Komagataeibacter hansenii*. Sulfate reducing bacteria (SRB), such as *Desulfomonile* sp., *Desulfotomaculum* sp. or *Desulfosarcina* sp., were not detected in the cap rock materials.
4. The Gorleben sediments yielded a negative result with respect to the determination of various enzyme activities (DMSO reduction, dehydrogenase activity, hydrolysis of fluorescein diacetate), in contrast to the control samples such as garden soil and compost.

5. The Gorleben sediment samples were also examined with respect to the biomass, which is also relevant for non-living bacteria. The analyses examined for the presence of biomarkers. The fatty acid patterns showed a quantity close to the detection limit. A suitable parameter would be the DNA content of the soil samples which were not available when the report was published.
6. It could not be ruled out that a reduction of the number of bacteria occurred by the deep freezing of the soil for the shaft drilling. For this reason, tests were performed. It could be shown that up to 450 days the freezing did not affect the autochthonous microorganisms, in contrast to parallel samples inoculated with pure cultures of different bacteria. The autochthonous bacteria population of the samples was constant and accounted to $2.4 \cdot 10^6$ CFU/g (CFU: Colony Forming Units).

## 5.2 Microbial effects on radionuclide migration

The sorption behavior of the redox sensitive radionuclides $^{95m}Tc$ and $^{75}Se$ was investigated with regard to microbial metabolism and the development of bacterial populations in loose sediments. In this project (funded by the Federal Ministry of Research and Technology) , soil samples collected at the Gorleben site (indicated by Gorleben and Trebel) were investigated and compared to a sample from the Berlin region [49]. Tab. IX shows the macroscopic chemical and physical characteristics of these soils and Fig. 10 shows the observed spectrum of microbial species found in the soils.

Tab. IX Chemical and physical characteristics of the soils under investigation

| | Gorleben sand soil | Trebel sand soil | Lankwitz glacial till |
|---|---|---|---|
| Sand [%] | 98.5 | 99.0 | 71.2 |
| Silt [%] | 1.5 | 10.0 | 25.3 |
| Clay [%] | 0 | 0 | 3.5 |
| specific surface [$m^2/g$] | 1.8 | 0.8 | 7.4 |
| CEC [meq/100 g] | 0.9 | 0.5 | 6.2 |
| total carbon content [%] | 0.01 | 0.005 | 0.1 |
| inorganic C [%] | 0.06 | 0.07 | 0.7 |
| organic C [%] | 0.05 | 0.064 | 0.5 |

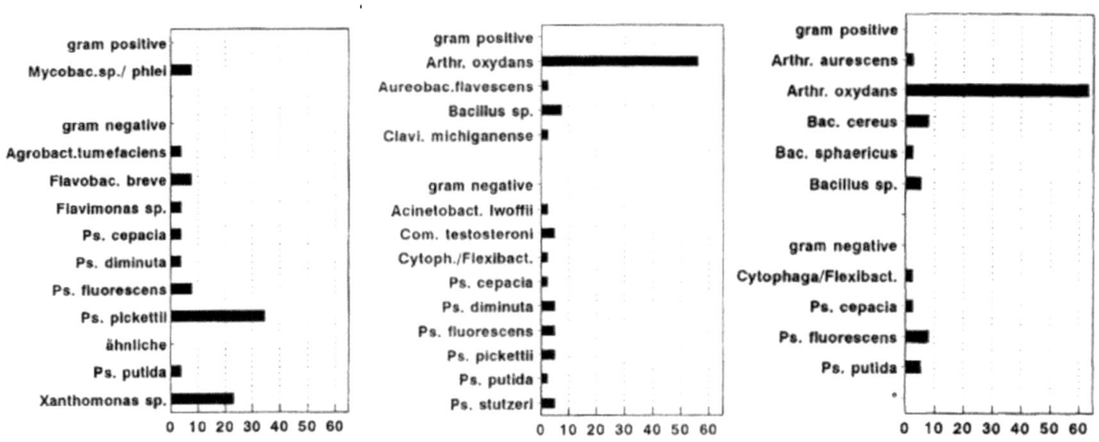


a) Gorleben sand | b) Trebel sand | c) Lankwitz till

Fig. 10 Spectrum of microbial species in the different soils (%).

The distribution of organisms shown in Fig. 10 suggests that the Gorleben sand might have more redox effects in comparison with the Trebel and Lankwitz soils. Batch experiments under aerobic and anaerobic conditions and recirculation column experiments (aerobic conditions) were carried out using sterile and non-sterile soil samples.

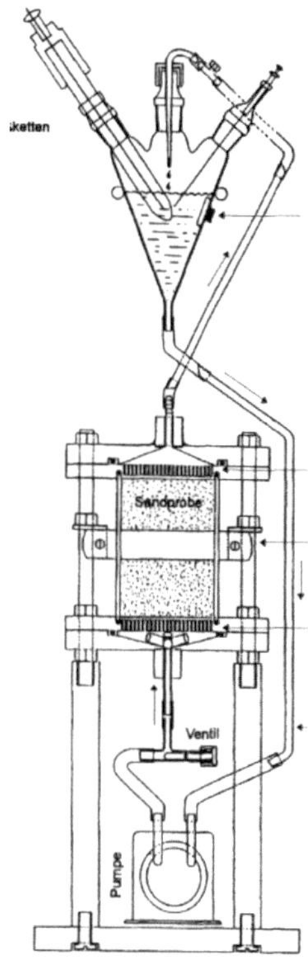


Fig. 11 Experimental set-up of recirculation columns for investigation of the microbial effects on radionuclide migration under sterile and non-sterile conditions.

Sterile migration experiments were compared with non-sterile experiments. The investigation of sterilizing methods with low impact on the physico chemical properties of sediments proved gamma irradiation to be the best choice. Other techniques under investigations include treatment with $NaN_3$, ethylene oxide, mercury(II) chloride or formaldehyde, as well as heat treatment, freezing and UV irradiation. Nutrients (200 ppm) were added to the batch experiments consisting of sodium acetate, nitrate and phosphate with the ratio C:N:P = 100:5:1. Following results were obtained:

- In batch experiments, the addition of nutrients resulted in an immobilization of Tc and Se combined with a decreasing redox value (Eh).
- Non-sterile recirculation experiments showed a reproducible fixation of Tc and Se without any observed decrease of the redox value.
- The immobilization occurred without any measurable alteration of the macro environment.
- In sterile column experiments, no fixation of Tc and Se occurred within the experimental duration (95 days).
- The addition of microorganisms isolated from the non-sterile columns led to a decreasing of the radionuclide concentration. This removal was not connected with decreasing redox values (*Aeromonas* sp., *Vibrio* sp. *Sphingomonas* sp., *Chryseomonas luteola* (now *Pseudomonas luteola*), *Sphingomonas paucimobilis*, and *Variovorax paradoxus*). Chryseomonas luteola is known to adsorb heavy metals and , and Xanthomonas sp. Present in the Gorleben sand can reduce selenite.
- The addition of biocide (5000 ppm $NaN_3$) to the non-sterile columns resulted in a remobilization of Tc but not of Se.
- To a great extent the microorganisms identified within the non-sterile columns were allochthonous (see chapter 5.3).
- Up-take of pertechnetate with simultaneous presence of reducing agents was detected for five bacterial strains (see above). The technetium could be fixed to cell fragments, dead biomass and living bacteria. The immobilization of Tc by living cells was found to be much stronger than by dead cells.

However, the statement that in sterile column experiments, no fixation of Tc and Se occurred within the experimental duration was not really confirmed by the retention coefficients reported in the final document [49]. In this report Rs values were given (see Tab. X).

Tab. X Sorption coefficients of Tc and Se in sand column experiments under sterile and non-sterile conditions with different ground water compositions.

| Condition | Solution Ionic strength mol/l | **Rs Tc** ml/g | **Rs Se** ml/g |
|---|---|---|---|
| Sterile | 0.75 | 0.0 – 0.54 | 0.0 – 0.60 |
| Sterile | 0.75 | 0.0 – 0.58 | 0.01 – 0.72 |
| Sterile | 0.016 | 0.0 – 0.90 | 0.03 – 0.65 |
| Sterile | 0.016 | 0.0 – 0.36 | 0.0 – 0.35 |
| non-sterile | 0.016 | 0.34 – 0.62 | 0.16 – 0.30 |
| non-sterile | 0.016 | 0.0 – 0.67 | 0.0 – 0.46 |
| non-sterile | 0.75 | 0.0 – 0.33 | 0.0 – 0.21 |
| non-sterile | 0.75 | 0.0 – 0.24 | 0.0 – 0.18 |

The authors of the reports [49-53] claimed that the batch experiments were performed under aerobic and anaerobic conditions, however, differences were not discussed. The redox potentials of the waters were measured in the range between +300 and +600 mV. As a conclusion of the report, it was stated that due to the existing microbial population in the cap rock aquifers and in the disposal, the cap rock aquifer's retardation capacity for radionuclide migration might be overestimated not knowing the impact of the autochthonous microflora on those radionuclides interacting with micro-organisms.

## 5.3 Microbial population in Gorleben groundwater – soil systems

At the Institute for Radiochemistry of the Technical University Munich, the migration of the fission product Tc and the actinides Np, Pu, Am was investigated in cap rock materials of the Gorleben salt dome. The processed rock samples were equilibrated with different groundwater collected at the Gorleben site. Some of these groundwaters showed a high concentration of organic carbon, mainly in the form of fulvic and humic substances. These organic macro molecules complex radionuclides preventing sorption onto the solids. They consist of heterogeneous mixtures of compounds, such as complex aromatic macromolecules with all kind of functional groups, including phenolic OH and COOH groups.

The process by which humic substances are formed is not well understood. It is assumed that they are residues of the metabolism of microorganisms. This was the reason why the microbial contamination of the sorption samples gained attention.

The rock samples were obtained during the drilling of shaft I. As described in chapter 5.1, the soil had been frozen for several years and the samples were kept at -20°C until they were processed. The cap rock samples covered sandy, till, argillaceous and clay materials from depths between -8 and -187.5 m below sea level. The corresponding groundwaters were sampled from nearby sampling wells.

The Gohy 1281 groundwater had especially high concentrations of fulvic and humic substances. This groundwater was investigated in details [54]. Following information was given (Tab)

Tab. XI Characteristics of the groundwater Gohy 1281

| | **pH** | **Eh**<br>mV | **Salt content**<br>mg/l | **DOC**<br>mg/l |
|---|---|---|---|---|
| **Gohy 1281** | 7.5 | +337 | 5428 | 10.6 |

As it was not possible to check the possibility of a microbial contamination during the on-site sampling, investigations were started concerning possible microbial contamination in the laboratory. Plate counts of following potential microbial contamination sources were determined:

- Air of the laboratory and atmosphere of the gloveboxes
- Vessels, such as the preservation vessels used for the conditioning of solids with groundwater
- Vials used for batch sorption experiments (20 ml)
- Filtration devices and filters.

Plate-Count-Agar was used as the growth medium. Details of sampling are described in reference [54]. The filtration device was sampled directly after filtration of the Gohy S106-1281 system and one week later. A sterile phosphate-buffered saline (PBS) solution was used. The plate counts were reported in CFU/l units. In additional to counting, the distribution of microbial species was analyzed with respect to the different sources of contamination. The species were distinguished according to spore formation, specific metabolisms and their origin. Anaerobic microorganisms were grouped in fungi, micro coccoid, typical soil and groundwater species (Arthrobacter sp., Azotobacter sp., Pseudomonas sp.), and endospore formers (Bacillus sp.). By means of enrichment culture techniques, sulfate, iron and manganese reducers as well as anaerobic endospore formers were investigated. Other species were not identified.

The number of cells in the humic-rich groundwater Gohy 1281 used for the sorption experiments was determined as follows:

- Aerobic microorganisms: $3.4x10^3$ CFU/l,
- anaerobic microorganisms: $1.2x10^3$ CFU/l.

82.3 % of the aerobic microbes and 91.4 % of the anaerobic cells were contaminants from the filtration device. The vessels contributed 7.4 % to the aerobic cell fraction and 3.3 % to the anaerobic species. The contribution of the filters accounted for 7.3 % (aerobic) and 3.1 % (anaerobic), whereas from the gloveboxes 3.0 % of the aerobic and 2.2 % of the anaerobic microorganism contaminated the groundwater.

(anaerobic), whereas from the gloveboxes 3.0 % of the aerobic and 2.2 % of the anaerobic microorganism contaminated the groundwater.

The population of microorganisms in the groundwater Gohy 1281 was 581 CFU/l of aerobic and 103 CFU/l of anaerobic species. For comparison, the US EPA's National Primary Drinking Water Standards [55] considers heterotrophic plate counts as a method of measuring the variety of bacteria present in water samples. The maximum contaminant level is 500 CFU/ml. This value is by a factor of 1000 above the values reported for Gohy 1281.

## 5.4 Preliminary Safety Analysis of Gorleben (vSG)

In the Gorleben salt dome, gaseous hydrocarbon compounds (mainly methane) are present in concentrations up to 0.3 g per 1000 kg rock salt. However, the hydrocarbons are not distributed homogenously, in part of one drill core between 20 and 100 g per 1000 kg rock salt were detected. Condensates consist of low temperature boiling hydrocarbons (C6 to C16 mainly < C10) and are composed of saturated (56 – 80%), aromatic (6 – 11%) and some asphaltenes [56, 57]. The hydrocarbons can promote the

- thermochemical sulfate reduction (TSR) and
- microbial sulfate reduction (BSR).

As TSR occurs only at temperatures above 80 to 130°C, this reaction is irrelevant for a low-level radioactive waste repository. BSR can take place in the low temperature regime. In the presence of water, the overall BSR reaction equation using methane as representative hydrocarbon compound is given in eq. (3).

$$CaSO_4 + CH_4 \rightarrow CaCO_3 + H_2S + H_2O \quad (3)$$

Reaction (3) results in the formation of water. In the scope of investigations of metal accumulation during and after deposition of the Kupferschiefer, BSR was elaborated [58].

Investigations of thin sections of drill cores taken from the Gorleben rock salt showed concentrations of liquid hydrocarbons in the range of 40 to 250 ml $m^{-3}$ in which most samples contained concentrations below 100 ml $m^{-3}$ [59, 60]. Furthermore, an average water content of the Gorleben rock salt of 0.02 vol.% was taken into account [61]. Using these data, the mechanical effects of thermal or microbial sulfate reduction were estimated [56], resulting in a maximum increase of volume of the rock salt by less than 1%. Investigations of rocksalt formations where TSR had taken place, showed additional porosity which was partially backfilled by carbonates and sulfides [62].

Dissimilatory $SO_4^{2-}$-reducing bacteria reduce $SO_4^{2-}$ to $H_2S$, according to the following simplified net mass balance reaction [63]:

$$2CH_2O + SO_4^{2-} \rightarrow 2HCO_3^- + H_2S + \text{biomass} \quad (4)$$

Upon bacterial $SO_4^{2-}$-reduction, metabolic residues released from other bacteria are used as a carbon source. At the same time, production of new biomass takes place due to the growing population of $SO_4^{2-}$-reducers. For this reason, organic matter of ancient sediments, in which $SO_4^{2-}$-reduction was the final step of organic matter decomposition, is still enriched in bacterial lipids. A scheme of the mechanisms of the BSR is shown in Fig. 12.

For a deep underground disposal, where only methane is available as organic material, the possibility of the sulfate reducing process is not obvious. In Schink's review in 1997 [16], it was stated that "anaerobic, sulfate-dependent methane oxidation appears to be an important redox reaction in anoxic marine sediments, but so far nobody can explain which bacteria catalyze this reaction and how they overcome the problem of activation of the very stable methane molecule."

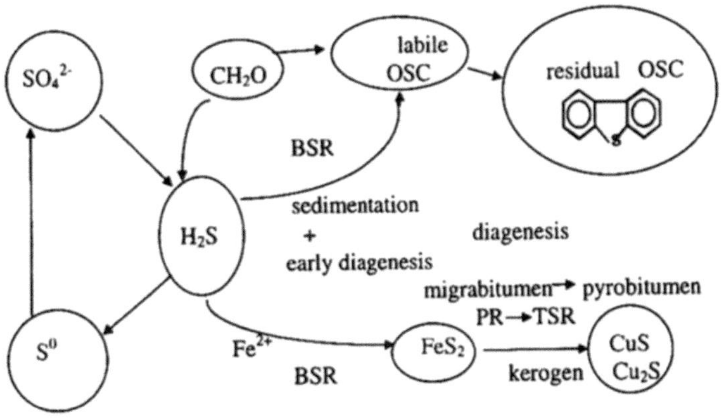


Fig. 12 Scheme of the mechanisms of the microbial sulfate reduction and the formation of dissolved organic ligands [58].

However, there might be other organics present in the disposal horizon that can be oxidized by sulfate-reducers without concomitant oxidation of methane. In Gorleben, the hydrocarbons could serve as substrates for sulfate reducing bacteria. It is not clear, if any exist within the rock salt itself, undoubtedly they exist in the overlying briny groundwaters.

Marchal [64] analyzed common BSR reaction products and the relevant amounts. One of his limiting factors to BSR was that $H_2S$ can be generated during BSR as long as sulfate and organic nutrients for the microbes are available, but only as long as the $H_2S$ concentration is below the level that is toxic for bacterial metabolism. Reis et al. [65] investigated cultures of sulfate reducing bacteria (SRB) growing on lactate and sulfate incubated at $5.8 \leq pH \leq 7.0$. The effect of pH on growth rate was determined. The highest growth rate was observed at pH 6.7. Hydrogen sulfide produced from sulfate reduction was found to have a direct and reversible toxicity effect on the SRB. A hydrogen sulfide concentration of 547 mg/L (16.1 mM) completely inhibited the culture growth.

# 6 Previous EU projects

Due to the laboratories and equipment of KIT-INE, our own research activities on the effects of microbial activities have not been pursued. However, KIT coordinated two European Framework Programmes where microbial processes took broad importance.

## 6.1 Funmig

The EURATOM EC 6th Framework Program Integrated Project "Fundamental Processes of Radionuclide Migration" (IP FUNMIG) was organized in workpackages attempting to cover the key migration processes at different scales (laboratory to repository) and in different geological media. The geological media focused mainly on clay and granite. A complete overview of the outcome of IP Funmig can be found in a special issue of Applied Geochemistry 27, 2, 2012 [66-77].Two media independent workpackages concerned laboratory studies of "well established" and with "less-established" migration processes. Within the study of the "less-established" migration processes, investigations were performed by several groups.

A workpackage comprised the role and extent of microbiologically mediated processes affecting radionuclide migration in the far-field of a repository. The focus was on the identification of radionuclide interaction with detached microorganisms, retention through interaction with biofilm material, including microbiologically mediated reduction, and change in the geochemical environment generated by microbial metabolism, including the generation and influence of various chelating compounds.

Cultivation of biofilms was started with the inoculation of standard medium with *Pseudomonas stutzeri*, a well-known biofilm producer. The experiments were performed in an air atmosphere under nonsterile conditions, and a multispecies biofilm developed [78, 79]. Sorption studies with Th(IV), Am(III), Co(II) and Cs(I) on *in-situ* grown biofilm and on apatite and quartz slides without biofilms were carried out. The biofilm was grown on mineral slides under *in-situ* conditions at the Äspö HRL. The mass of biofilm on the surfaces was approx.-imately the same on the different minerals, corresponding to 1.25 ± 0.13 million cells $cm^{-2}$. By time dependent sampling and analysis, the sorption kinetics onto the slides could be determined. Autoradiography analysis was used to localize where sorption took place on the mineral slides.

The interactions of U with biofilms were investigated and it was observed that U immobilization was associated with extracellular polymeric substances and not with cell membranes or bacterial cells. Electrochemical micro sensor studies of the $O_2$ concentrations within the biofilm identified depleted zones close to the interface between biofilm and air which might trigger uranium redox processes.

Another study focused on the role of iron oxides in retarding U transport by acting as a sorbent for uranyl. Reductive dissolution of iron oxides by S(-II), the product of microbial sulfate reduction, can cause the mobilization of adsorbed uranium. However, reduction of U(VI) by S(-II) could counteract this effect due to the formation of U(IV) and subsequent precipitation of $UO_2$. The following conclusion was drawn from the results of the measurements: The release of S(-II) in subsurface environments could lead to the mobilization of U(VI) which was adsorbed to iron oxides.

## 6.2 ReCosy

One workpackage of the EURATOM EC 7$^{th}$ Framework Program Collaborative Project "Redox Phenomena Controlling Systems (CP ReCosy) was entitled "Redox Reactions of Radionuclides" [80]. This workpackage was divided in two parts:

- Chemical and redox behavior of the investigated radionuclides in the different systems.
- Chemical and redox behavior of the investigated radionuclides in the different systems through microbially mediated processes.

The Research Center Rossendorf and the Institute of Physics, Lithuania participated within the second part of the workpackage. The complete list of publications on this topic can be found in the summary paper by Duro [80]. The work was focused on the study of the microbial impact and on the uranium redox state of *in-situ* biofilms with emphasis on biologically mediated redox processes.

Biofilms were composed of bacteria, fungi, algae, protozoa, exopolymeric substances (EPS), corrosion products and 50 - 95% water. Biofilms are ubiquitous and have to be considered as an important factor in natural biogeochemical processes influencing the redox state of radionuclides. They show a multiplicity of interactions with metals and contribute to metal mobility or immobilization. To investigate the properties of the biofilms an improved miniaturized oxygen micro sensor and electrochemical micro sensors were developed which allowed the application in several URLs.

Dedicated studies were performed to investigate the influence of microorganisms on the redox behavior of Pu and Tc. The results indicated that *Bacillus mycoides* was able to reduce the initial Pu(IV) to Pu(III) at low pH. Experiments into the combined effect of microorganisms and iron-bearing mineral hematite on Tc(VII) sorption showed that after contact of Tc(VII) for 144 h with bacteria *Arthrobacter globiformis* and *Cellulomonas cellulans* 78 and 98% of Tc remained in solution. Differences in Tc(VII) sorption processes onto hematite due to microbial activity of microorganisms isolated from the groundwater borehole were observed as well.

Further experiments with only microorganisms and without Fe-oxides were performed. The results indicated that the sorption of Tc(VII) by hematite under neutral or slightly alkaline pH values was due to the presence of microorganisms that could reduce Tc(VII) to less mobile forms of Tc(IV), but that in the absence of microorganisms no important sorption of Tc(VII) by hematite was observed.

Other European Projects such as NF-Pro or Pamina covered microbial activities only with respect to their $CO_2$ gas production by degradation of organic waste components. Detailed investigations have not been carried out in these projects.

# 7 Conclusions

Microorganisms might give rise to reactions and a release of radionuclides and their transport in the geosphere in the long term. In the scope of German safety cases, the existence of microbial effects was not ignored, but considered of minor priority. Tab. XII gives an overview of the number of bacteria in deep groundwater environments relevant for disposal. The populations found in the cap rocks of the Gorleben site correspond well with findings at sites in granite host rocks. In clays/argillaceous host rocks, however, the number of colony forming units was smaller. Data were not determined for the rock salt at the Asse II salt mine and Gorleben.

Tab. XII Bacterial populations in deep groundwater environments, in the Gorleben cap rock and in the Waste Isolation Pilot Plant (WIPP).

| Location | Geology | Depth / m (below ground) | Bacterial Count | Ref. |
|---|---|---|---|---|
| Canada | Granite | 350 - 400 | Total counts $10^3$ - $10^5$ cells $ml^{-1}$ | [81] |
| Finland | Granite (various) | 200 - 950 | $10^5$ - $10^6$ cells $ml^{-1}$ (includes fresh, brackish and saline waters) | [81] |
| Japan | Granite | 400 - 790 | Total counts $10^2$ - $10^7$ bacteria $ml^{-1}$ | [81] |
| Stripa, Sweden | Granite | 799 - 1240 | Total counts $10^5$ cells $ml^{-1}$ | [81] |
| Aspo, Sweden | Granite | 129 - 1078 | Total counts $10^6$ cells $ml^{-1}$ | [81] |
| Grimsel, Switzerland | Granite | Approx. 350 | $10^5$ CFU $ml^{-1}$ | [81] |
| Altnabreac, UK | Granite | 10-281 | $10^6$ CFU $ml^{-1}$ | [81] |
| Mol, Belgium | Boom Clay | 190 - 223 | $10^3$ CFU $ml^{-1}$ | [81] |
| Harwell, UK | Oxford Clay | 165-331 | $10^4$ - $10^5$ CFU $ml^{-1}$ | [81] |
| Asse, Germany | Salt | 750 | Not determined | |
| Yucca Mtn., USA | Volcanic Tuff | 60 | $10^2$ - $10^3$ bacteria $g^{-1}$ dry weight | [81] |
| Gorleben, Germany | Sediment cap rock | 115 - 170 | $2.4 \cdot 10^6$ CFU $g^{-1}$ | [47, 48] |
| Culebra (overlying WIPP/USA) | fractured dolomite | 213 - 250 | 2.4-8.6 x $10^5$ cells $ml^{-1}$ | [82] |
| WIPP halite | Salado formation | | $10^4$ CFU/g | [83] |

In the Appendix (Tab. XIII), the types of microorganisms under investigation are listed with respect to the degradation of cellulose, bitumen, compacted wastes, and cement as well as the species found in the Gorleben cap rocks. The table shows relatively few overlaps, only the bacteria *Arthrobacter* was tested in bitumen and the compacted waste investigations, *Clostridium sporogenes* in cellulose and compacted wastes test. The fungus *Aspergillus niger* was analyzed in the compacted wastes experiments and was found in the Gorleben cap rock samples.

According to Tittel et al. [48], the following processes and consequences need to be discussed in the safety case:

Processes:

- degradation of the waste forms
- corrosion of metallic materials
- alteration of the backfill material
- alteration of the chemical composition of waters, e.g. Eh and pH
- generation of complexing agents by degradation of organic material
- uptake of radionuclides by microorganisms.

Consequences:

- gas production ($CO_2$, $H_2$, $CH_4$) and formation of explosive gas/air mixtures
- release of volatile radionuclides
- increase in pressure
- accelerated release of radionuclides
- increased solubility of radionuclides
- changes in the transport of radionuclides through the geosphere

Even if alteration of the Eh and pH of water was considered as an important effect, no measurements of the redox potentials were reported. The formation and concentrations of microbially generated complexing agents have never been quantified.

The following conclusions concerning the processes and consequences can be drawn:

- The conclusions on microbial degradation of paper, plastics and textiles are based on experiments performed in the context of WIPP. The gas production was determined and a gas production rate calculated. Gillow [31] reported a $CO_2$ production in the case of anaerobic experiments of $6.4x10^{-3}$, $1.4x10^{-2}$, and $5.6x10^{-2}$ µmol $CO_2$ $g^{-1}$ cellulose $day^{-1}$. These numbers took into account the whole $CO_2$ formation from the beginning of the experiments. Considering the long-term part Gillow's measurements, a lower production rate of $6.4x10^{-3}$ µmol $g^{-1}$ $day^{-1}$ for samples without nutrients and $1.4x10^{-2}$ µmol $g^{-1}$ $day^{-1}$ for samples with excess nitrate was obtained. These numbers were not biased by consumption of nutrients and nitrate.
- The microbial degradation processes of bitumen are also based on American investigations in the context of WIPP. As bituminized wastes are minor components in Germany, the radionuclide release by leaching was taken into account. Microbial

processes were considered only for predicting an underground sealing in the ERAM facility. A penetration of a degradation front of 6 µm $yr^{-1}$ was assumed.

- Microbial degradation of compacted mixed wastes was investigated. The experiments were performed at the University Darmstadt. The outcome of the tests showed that at 50°C none of the tested microorganisms survived longer than one month. Some aerobic spore formers survived at 28°C at low humidity, as well as filamentous fungi and yeasts. Microorganisms collected from mold sludge were found to be resistant against environmental disturbances.
- Microbial attack of cement and other backfill materials was not taken into account in any Safety Case in Germany.
- Microbial sulfate reduction (BSR) was considered in the preliminary safety analysis of Gorleben. In the case of a high-level waste disposal, where huge amounts of steel are present, the corrosion processes produce hydrogen up to high pressures. Based on literature data and the high $H_2$ partial pressure, it can be assumed that microbial sulfate reduction by degradation of hydrocarbon compounds (present in the Gorleben salt dome) is of minor importance to the safety of the disposal.

Due to the fact that in the former German disposal concepts fine grained carbon steel was a constituent of the canisters for LLW as well as for spent nuclear fuel disposal casks. Under the expected conditions in a rock salt disposal, the material showed even under anaerobic conditions high corrosion rates leading to the formation of hydrogen gas. Due to the high corrosion rates, microbial processes were assumed to be of less relevance.

The alteration of the chemical composition of waters, e.g. Eh and pH is influenced by the presence of cement, backfill materials and the hydrogen formed by iron corrosion. Microbial activities have to adapt to the environmental conditions. Especially in a rock salt disposal where the conditions for microbial life are unfavorable, alteration of the chemical composition of waters by microorganisms can be excluded, even if some microorganisms could induce precipitation of salt minerals. In the case of penetration of magnesium-rich brine into the disposal, a window-of-life could be determined. The water activity of a so called "Q-brine" would not allow for cell metabolism.

The cap rocks above the Gorleben salt dome were analyzed with respect to its microbial population (University Darmstadt). In the cap rock, ground water with high concentrations of fulvic and humic substances were found. A huge set of data exist concerning the effect of these substances on the complexation of radionuclides and on their migration behavior. Studies on the effect of microbial activities on the retention behavior of the redox-sensitive elements technetium and selenium have been performed. The experiments using sandy soils from the Gorleben region did not show significant changes in the sorption coefficients in sterile and under non-sterile conditions.

In the past, investigations of microbial activities and their effects on radionuclide retention in the deep underground disposal in Germany have not been performed comprehensively.

However, the effects were not ignored, some experimental investigations were carried out and the international literature was considered.

Due to the huge variety of microorganisms, it cannot be ruled out that microbial life and metabolisms exist in a deep geological nuclear waste disposal. However, the conditions for microbial metabolism are extremely unfavorable.

- In rock salt, the high ionic strength prevents the metabolism of many microorganisms (analogue: curing of meat). Only extreme halophiles can grow in NaCl solutions of 5 mol/l.
- In the presence of cement, the pH rises to values of > 12 (OPC) or about 11 (low pH cement). In these ranges, few microorganisms prefer to live.

# 8 References


[1] L. Abrahamsen, T. Arnold, H. Brinkmann, N. Leys, M.Merroun, K. Mijnendonckx, H. Moll, P.Polvika, A. Ševců, J. Small, M. Vikman, and K. Wouters, "A review of anthropogenic organic wastes and their degradation behaviour," Microbiology in Nuclear Waste Disposal (MIND), Deliverable 1.1, 2015.

[2] R. Knauer, "MIKROBEN: Gut leben im Endlager." *Spektrum der Wissenschaft,* 2016.

[3] L. Hallbeck, "Principal organic materials in a repository for spent nuclear fuel," Svensk Kärnbränslehantering AB, Stockholm, TR-10-19, 2010.

[4] ANDRA, "Dossier 2005 Argile, Tome: Évolution phénoménologique du stockage géologique," DOSA_269_TEP.pdf, 2005.

[5] J. West, "A review of progress in the geomicrobiology of radioactive waste disposal." *Radioactive Waste Management and Environmental Restoration,* Vol. 19, pp. 263-283, 1995.

[6] J. M. West and I. G. McKinley, "The Geomicrobiology of Nuclear Waste Disposal." *MRS Proceedings,* Vol. 26, 1983.

[7] R. Köster, "Zusammenstellung der vom Arbeitskreis erarbeiteten Vorschläge zur Charakterisierung, Qualitätssicherung und Datensammlung von LAW/MAW-Produkten," INE, Karlsruhe, 1982.

[8] IAEA, "Management of radioactive waste from nuclear power plants," International Atomic Energy Agency, Vienna, IAEA-TECDOC-276, 1981.

[9] B. Buchheim, H. Meyer, and M. Tholen. (2004). *Bestimmung des Inventars an chemischen und chemotoxischen Stoffen in den eingelagerten radioaktiven Abfällen der Schachtanlage Asse*. Available: http://www.asse.bund.de/SharedDocs/Downloads/Asse/DE/IP/historische-dokumente/stilllegungskonzept-hmgu/10-chemische-chemotoxische-stoffe.html

[10] F. Peiffer, B. McStocker, D. Gründler, F. Ewig, B. Thomauske, A. Havenith, and J. Kettler, "Abfallspezifikation und Mengengerüst: Basis Ausstieg aus der Kernenergienutzung," Gesellschaft für Anlagen- und Reaktorsicherheit (GRS) mbH, Köln, GRS - 278, Vorläufige Sicherheitsanalyse für den Standort Gorleben, 2011.

[11] K. Pedersen, "The deep biosphere." *GFF,* Vol. 132, pp. 93-94, 2010.

[12] M. T. Madigan, J. M. Martinko, and T. D. Brock, *Brock biology of microorganisms*. Upper Saddle River, NJ: Pearson Prentice Hall, 2006.

[13] T. J. DiChristina, J. K. Fredrickson, and J. M. Zachara, "Enzymology of electron transport: Energy generation with geochemical consequences," in *Reviews in Mineralogy & Geochemistry: Molecular Geobiology*. Vol. 59, J. J. Rosso, Ed., ed Chantilly, VA: Mineralogical Society of America, pp. 27-52, 2005.

[14] A. Kappler and K. L. Straub, "Geomicrobiological cycling of iron," in *Reviews in Mineralogy & Geochemistry: Molecular Geobiology*. Vol. 59, J. J. Rosso, Ed., ed Chantilly, VA: Mineralogical Society of America, pp. 85-108, 2005.

[15] R.A.Bruce, L. A. Achenbach, and J. D. Coates, " Reduction of (per)chlorate by a novel organism isolated from paper mill waste." *Environmental Microbiology* Vol. 1, pp. 319-329, 1999.

[16] B. Schink, "Energetics of Syntrophic Cooperation in Methanogenic Degradation." *MICROBIOLOGY AND MOLECULAR BIOLOGY REVIEWS,* Vol. 61, pp. 262-280, 1997.

[17] M. Egger, M. Hagens, C. J. Sapart, N. Dijkstra, Niels A.G.M. van Helmond, J. M. Mogolón, N. Risgaard-Petersen, Carina van der Veen, S. Kasten, N. Riedinger, M. E. Böttcher, T. Röckmann, Bo Barker Jørgensen, and C. P. Slomp, "Iron oxide reduction in methane-rich deep Baltic Sea sediments." *Geochimica et Cosmochimica Acta,* Vol. 207 pp. 256-276, 2017.

[18] M. Egger, O. Rasigraf, C. J. Sapart, T. Jilbert, M. S. M. Jetten, T. Röckmann, C. van der Veen, N. Bândă, B. Kartal, K. F. Ettwig, and C. P. Slomp, "Iron-Mediated Anaerobic Oxidation of Methane in Brackish Coastal Sediments." *Environmental Science & Technology,* Vol. 49, pp. 277-283, 2015.

[19] D. R. Wessner, C. Dupont, and T. Charles, *Microbiology*. Hoboken, NJ, Wiley, 2013.

[20] J. E. Hallsworth, M. M. Yakimov, P. N. Golyshin, J. L. M. Gillion, G. D'Auria, F. De Lima Alves, V. La Cono, M. Genovese, B. A. McKew, S. L. Hayes, G. Harris, L. Giuliano, K. N. Timmis, and T. J. McGenity, "Limits of life in $MgCl_2$-containing environments: chaotropicity defines the window." *Environmental Microbiology,* Vol. 9, pp. 801-813, 2007.

[21] M. M. Yakimov, V. L. Cono, G. L. Spada, G. Bortoluzzi, E. Messina, F. Smedile, E. Arcadi, M. Borghini, M. Ferrer, P. Schmitt-Kopplin, N. Hertkorn, J. A. Cray, J. E. Hallsworth, P. N. Golyshin, and L. Giuliano, "Microbial community of the deep-sea brine Lake Kryos seawater–brine interface is active below the chaotropicity limit of life as revealed by recovery of mRNA." *Environmental Microbiology,* Vol. 17, pp. 364-382, 2015.

[22] A. Stevenson, J. Burkhardt, C. S. Cockell, J. A. Cray, J. Dijksterhuis, M. Fox-Powell, T. P. Kee, G. Kminek, T. J. McGenity, K. N. Timmis, D. J. Timson, M. A. Voytek, F. Westall, M. M. Yakimov, and J. E. Hallsworth, "Multiplication of microbes below 0.690 water activity: implications for terrestrial and extraterrestrial life." *Environmental Microbiology,* Vol. 17, pp. 257-277, 2015.

[23] J. A. Cray, J. T. Russell, D. J. Timson, R. S. Singhal, and J. E. Hallsworth, "A universal measure of chaotropicity and kosmotropicity." *Environmental Microbiology,* Vol. 15, pp. 287-296, 2013.

[24] R. Storck, "Langzeitsicherheitsaspekte der Endlagerung von Transuranelementen," GSF-Bericht 15/92, 1992.

[25] W. Levi, G. Memmert, H. Venzlaff, and H.-J. Wingender, "Projekt Sicherheitsstudien Entsorgung (PSE): Zusammenfassender Zwischenbericht," Berlin, 1981.

[26] B. Kienzler, R. Köster, and E. Korthaus, "Berechnungen und Sensitivitätsbetrachtungen zur Aktivitätsfreisetzung aus einem ersoffenen Endlager für radioaktive Abfälle anhand eines einfachen Störfallmodells," Kernforschungszentrum Karlsruhe, KfK 3013, 1980.

[27] R. Storck, "Performance assessment of geological isolation systems for radioactive waste (PAGIS): Disposal in salt formations," Commission of the European Community, Brüssel, EUR 11778 EN, 1988.

[28] W. Bechthold, W. Braun, C. Brückner, K. D. Closs, U. Knapp, and R. Papp, "Systemanalyse Mischkonzept (Hauptband)," KfK AE KWA 2190 A1, 1989.

[29] A. J. Francis, J. B. Gillow, and M. R. Giles, "Microbial gas generation under expected Waste Isolation Pilot Plant repository conditions," Sandia National Laboratories (ANL), Aluquerque, NM, USA, SAND96-2582, 1997.

[30] J. B. Gillow and A. J. Francis, "3.1 Re-evaluation of microbial gas generation under expected waste isolation pilot plant conditions data summary and progress report (February 1 - July 15, 2002)," Environmental Sciences Department, Brookhaven National Laboratory, Upton, NY USA, 2002.

[31] J. B. Gillow and A. J. Francis, "Microbial gas generation under expected Waste Isolation Pilot Plant repository conditions: Final report," Brookhaven National Laboratory, Upton, NY, USA, BNL-96148-2011-IR, 2006.

[32] C. E. ZoBell and M. A. Molecke, "Survey of microbial degradation of asphalts with notes on relationship to nuclear waste management," Sandia Laboratories, Albuquerque, NM, SAND 78-1371, 1978.

[33] N. Ait-Langomazino, R. Sellier, G. Jouquet, and M. Trescinski, "Microbial degradation of bitumen." *Experientia,* Vol. 47, pp. 533-539, 1991.

[34] M. Wolf, "Mikrobieller Abbau von Bitumen.," NAGRA TB 89-14, 1989.

[35] M. Wolf and R. Bachofen, "Microbial degradation of bitumen." *Experientia,* Vol. 47, pp. 542-548, 1991.

[36] H. Rauche, P. Sitz, V. Lukas, K. Rumphorst, G. Lippmann, K. Wagner, and T. Teichert, "Nachweisführungen zur Langzeitsicherheit, zur Tragfähigkeit und zur Gebrauchstauglichkeit der Sehachtverschlüsse ERA Morsleben," Bundesamt für Strahlenschutz (BfS), Salzgitter, P 177, Planfeststellungsverfahren zur Stilllegung des Endlagers für radioaktive Abfälle Morsleben, 2003.

[37] U. Krakau and V. Köckritz, "Langzeitbeständigkeit von Bitumen und Asphalten aus geologischer und historischer Sicht," in *Natürliche Analoga zur Endlagerung Radioaktiver Abfälle,* Kernforschungszentrum Karlsruhe, 4. und 5. November 1993, pp. 210-211, 1993.

[38] M. F. Libert, R. Sellier, G. Jouquet, M. Trescinski, and H. Spor, "Effects of Microorganisms Growth on the Long-Term Stability of Cement and Bitumen." *MRS Proceedings,* Vol. 294, 1992.

[39] H. J. Kutzner and D. Dengler, "Orientierende Untersuchungen über die Wahrscheinlichkeit mikrobieller Aktivitäten in gepreßten Mischabfällen unter den Deponiebedingungen der Grube Konrad," TU Darmstadt, im Auftrag von Nukem GmbH, Hanau, unveröffentlicht, 1986.

[40] B. Christ, "Complying with the conditions for ulitmate storage in Konrad by industrial process qualification." *Kerntechnik,* Vol. 51, pp. 108-112, 1987.

[41] A. Schneider, "Neue Diagramme zur Bestimmung der raltiven Luftfeuchtigkeit übersättigten wässerigen Salzlösungen und wässerigen Schwefelsäureldsungen bei verschiedenen Temperaturen." *HOLZ als Roh- und Werkstoff,* Vol. 18, pp. 269-272, 1960.

[42] R. D. Rogers, M. A. Hamilton, and J. W. McConnell, "The Possibility for Microbially Influenced Degradation of Cement Solidified Low-Level Radioactive Waste Forms." *MRS Proceedings,* Vol. 294, 1992.

[43] J. S. Swanson, A. Cherkouk, T. Arnold, A. Meleshyn, and D. T. Reed, "The Microbiology of Subsurface, Salt-Based Nuclear Waste Repositories: Using Microbial Ecology, Bioenergetics, and Projected Conditions to Help Predict Microbial Effects on Repository Performance," LA-UR-16-28895, 2016.

[44] A. Stevenson, J. A. Cray, J. P. Williams, R. Santos, R. Sahay, N. Neuenkirchen, C. D. McClure, I. R. Grant, J. D. R. Houghton, J. P. Quinn, D. J. Timson, S. V. Patil, R. S. Singhal, J. Anton, J. Dijksterhuis, A. D. Hocking, B. Lievens, D. E. N. Rangel, M. A. Voytek, N. Gunde-Cimerman, A. Oren, K. N. Timmis, T. J. McGenity, and J. E. Hallsworth, "Is there a common water-activity limit for the three domains of life?" *ISME J,* Vol. 9, pp. 1333-1351, 2015.

[45] B. Kienzler, C. Borkel, V. Metz, and M. Schlieker, "Long-Term Interactions of Full-Scale Cemented Waste Simulates with Salt Brines," Karlsruhe Institute of Technology, KIT SR 7721, 2016.

[46] A. Hollmann and G. Tittel, "Nuklidmigration im Deckgebirge des Endlagerortes Gorleben: Vorträge eines Fachgesprächs zu Fragen der Radionuklidmigration und Langzeitsicherheit des geplanten Endlagers Gorleben am 3. Juni 1992 im BfS," Fachbereich Nukleare Entsorgung und Transport, Bundesamt für Strahlenschutz, Salzgitter, BfS-ET-18/93, 1993.

[47] H. J. Kutzner, H. Sonnen, and F. Bachmann, "Microbiological investigation of soil samples of the caprock above the Gorleben salt dome," Salzgitter, Germany, Radionuclide migration in the caprock above the Gorleben repository, 1993.

[48] G. Tittel, H. Kutzner, D. Herrmann, H. Eschrich, and E. Warnecke, "Potential influence of microbial activity on the migration of radionuclides from repositories for radioactive wastes." *Radiochimica Acta,* Vol. 52, pp. 305-310, 1991.

[49] I. Stroetmann, P. Kämpfer, J. Schüle, R. Sokotnejad, C. Merz, and A. Winkler, "Sorptionsexperimente zur Beeinflussung der Radionuklidmigration durch mikrobielle Aktivitäten am Beispiel des Technetiums und Selen & Untersuchung zur Methodik von Sterilisationsverfahren sowie mikrobielle Überwachung von Sorptionsexperimenten," FG Hygiene, TU Berlin & FR Umweltgeologie, FU Berlin, 1993.

[50] G. Maue, I. Stroetmann, T. Taute, and A. Winkler, "Einfluss der autochtonen Mikroflora auf die Sorption und Remobilisierung des Technetiums und Selens (Abschlussbericht des gemeinsamen Forschungsvorhabens)," FG Hygiene, TU Berlin & FR Umweltgeologie, FU Berlin, 1996.

[51] G. Maue, I. Stroetmann, T. Taute, and A. Winkler, "Einfluss der autochtonen Mikroflora auf die Sorption und Remobilisierung des Technetiums und Selens an verschiedenen Sedimenten," FG Hygiene, TU Berlin & FR Umweltgeologie, FU Berlin, 1997.

[52] A. Winkler, "Untersuchungen zur Mobilität von Technetium (und Selen) in Norddeutschen Grundwasserleitern und Technetium im Kontakt mit natürlich vorkommenden Mineralien," Berliner Geowissenschaftliche Abhandlungen, Reihe A, Band 117, FU – TU – TFH Berlin, 1989.

[53] I. Stroetmann, "Einfluss mikrobieller Aktivität auf das Migrationsverhalten redoxsensitiver Radionuklide (Technetium und Selen) in Lockergesteinen," Fachgebiet Hygiene, Technische und Freie Universität Berlin, D 83, 1995.

[54] J. I. Kim, J. Heitz, C. Konrad, I. Pashalidis, W. Runde, P. Zeh, B. Kornprobst, and C. Stöwer, "Nuklidmigration (Tc, Np, Pu, Am) im Deckgebirge des Endlagerortes Gorleben," TU München - Institut für Radiochemie, 1994.

[55] EPA, "National Primary Drinking Water Standards," Environmental Protection Agency, Washington, DC, USA, EPA 816-F-01-007, 2001.

[56] J. R. Weber, J. Hammer, and O. Schulze, "Empfehlungen der BGR zur Berücksichtigung der Kohlenwasserstoff-Vorkommen im Hauptsalz des Salzstockes Gorleben im Rahmen einer vorläufigen Sicherheitsanalyse," Bundesanstalt für Geowissenschaften und Rohstoffe (BGR), Hannover, Projekt Gorleben GE442800000, 2011.

[57] G. Bracke, T. Popp, W. Püttmann, B. Kienzler, A. Lommerzheim, and H. C. Moog, "Berücksichtigung der Kohlenwasserstoffvorkommen in Gorleben: Bericht der Arbeitsgruppe Kohlenwasserstoffe," Gesellschaft für Anlagen- und Reaktorsicherheit (GRS) mbH, Köln, GRS - 285, Vorläufige Sicherheitsanalyse für den Standort Gorleben, 2012.

[58] Y. Sun and W. Püttmann, "Metal accumulation during and after deposition of the Kupferschiefer from the Sangerhausen Basin, Germany." *Applied Geochemistry,* Vol. 12, pp. 577-592, 1997.

[59] T. Popp, M. Fischer, and H. Kern, "Gefügekundliche Untersuchungen zur Verteilung von Kohlenwasserstoffen (KW) im Steinsalz." *Meyniana,* Vol. 54, pp. 131-154, 2002.

[60] M. Fischer, "Gefügekundliche Untersuchung von Steinsalz im Hinblick auf die Verteilung und Bindungsform von Kohlenwasserstoffen und die Bestimmung des Volumenanteils der Kohlenwasserstoff-Einschlüsse," Diploma Thesis, Institut für Geowissenschaften, Christian-Albrechts-Universität, Kiel, 2000.

[61] O. Bornemann, J. Behlau, R. Fischbeck, J. Hammer, W. Jaritz, S. Keller, G. Mingerzahn, and M. Schramm, *Projekt Gorleben, Standortbeschreibung Gorleben. Teil 3: Ergebnisse der über- und untertägigen Erkundung des Salinars* Vol. C 73. Hannover: Bundesanstalt für Geowissenschaften und Rohstoffe (BGR), 2008.

[62] W. Püttmann, "Literaturrecherche zur thermochemischen Sulfatreduktion (TSR)," Universität Frankfurt, 2012.

[63] H. G. Machel, H. R. Krouse, and R. Sassen, "Products and distinguishing criteria of bacterial and thermochemical sulfate reduction." *Applied Geochemistry,* Vol. 10, pp. 373-389, 1995.

[64] H. G. Machel, "Bacterial and thermochemical sulfate reduction in diagenetic settings – old and new insights." *Sedimentary Geology,* Vol. 140, pp. 143-175, 2001.

[65] M. A. M. Reis, J. S. Almeida, P. C. Lemos, and M. J. T. Carrondo, "Effect of hydrogen sulfide on growth of sulfate reducing bacteria." *Biotechnology and Bioengineering,* Vol. 40, pp. 593-600, 1992.

[66] L. Duro, "Preface." *Applied Geochemistry,* Vol. 27, p. 377, 2012.

[67] S. Altmann, C. Tournassat, F. Goutelard, J.-C. Parneix, T. Gimmi, and N. Maes, "Diffusion-driven transport in clayrock formations." *Applied Geochemistry,* Vol. 27, pp. 463-478, 2012.

[68] N. D. Bryan, L. Abrahamsen, N. Evans, P. Warwick, G. Buckau, L. Weng, and W. H. Van Riemsdijk, "The effects of humic substances on the transport of radionuclides: Recent improvements in the prediction of behaviour and the understanding of mechanisms." *Applied Geochemistry,* Vol. 27, pp. 378-389, 2012.

[69] J. Bruno and V. Montoya, "From aqueous solution to solid solutions: A process oriented review of the work performed within the FUNMIG project." *Applied Geochemistry,* Vol. 27, pp. 444-452, 2012.

[70] B. Schwyn, P. Wersin, J. Rüedi, J. Schneider, S. Altmann, T. Missana, and U. Noseck, "FUNMIG Integrated Project results and conclusions from a safety case perspective." *Applied Geochemistry,* Vol. 27, pp. 501-515, 2012.

[71] T. Behrends, E. Krawczyk-Bärsch, and T. Arnold, "Implementation of microbial processes in the performance assessment of spent nuclear fuel repositories." *Applied Geochemistry,* Vol. 27, pp. 453-462, 2012.

[72] L. Martinez-Landa, J. Carrera, M. Dentz, D. Fernàndez-Garcia, A. Nardí, and M. W. Saaltink, "Mixing induced reactive transport in fractured crystalline rocks." *Applied Geochemistry,* Vol. 27, pp. 479-489, 2012.

[73] T. Schäfer, F. Huber, H. Seher, T. Missana, U. Alonso, M. Kumke, S. Eidner, F. Claret, and F. Enzmann, "Nanoparticles and their influence on radionuclide mobility in deep geological formations." *Applied Geochemistry,* Vol. 27, pp. 390-403, 2012.

[74] U. Noseck, E.-L. Tullborg, J. Suksi, M. Laaksoharju, V. Havlová, M. A. Denecke, and G. Buckau, "Real system analyses/natural analogues." *Applied Geochemistry,* Vol. 27, pp. 490-500, 2012.

[75] C. Bruggeman, N. Maes, B. C. Christiansen, S. L. S. Stipp, E. Breynaert, A. Maes, S. Regenspurg, M. E. Malström, X. Liu, B. Grambow, and T. Schäfer, "Redox-active phases

and radionuclide equilibrium valence state in subsurface environments – New insights from 6th EC FP IP FUNMIG." *Applied Geochemistry,* Vol. 27, pp. 404-413, 2012.

[76] J. Lützenkirchen, "Summary of studies on (ad)sorption as a "well-established" process within FUNMIG activities." *Applied Geochemistry,* Vol. 27, pp. 427-443, 2012.

[77] P. E. Reiller, T. Vercouter, L. Duro, and C. Ekberg, "Thermodynamic data provided through the FUNMIG project: Analyses and prospective." *Applied Geochemistry,* Vol. 27, pp. 414-426, 2012.

[78] K. Groβmann, T. Arnold, E. Krawczyk-Bärsch, S. Diessner, A. Wobus, G. Bernhard, and R. Krawietz, "Identification of Fluorescent U(V) and U(VI) Microparticles in a Multispecies Biofilm by Confocal Laser Scanning Microscopy and Fluorescence Spectroscopy." *Environmental Science & Technology,* Vol. 41, pp. 6498-6504, 2007.

[79] E. Krawczyk-Bärsch, K. Grossmann, T. Arnold, S. Hofmann, and A. Wobus, "Influence of uranium (VI) on the metabolic activity of stable multispecies biofilms studied by oxygen microsensors and fluorescence microscopy." *Geochimica et Cosmochimica Acta,* Vol. 72, pp. 5251-5265, 2008.

[80] L. Duro, J. Bruno, M. Grivé, V. Montoya, B. Kienzler, M. Altmaier, and G. Buckau, "Redox processes in the safety case of deep geological repositories of radioactive wastes. Contribution of the European RECOSY Collaborative Project." *Applied Geochemistry,* Vol. 49, pp. 206-217, 2014.

[81] J. M. West and I. G. McKinley, "Progress in the Geomicrobiology of Radioactive Waste Disposal." *MRS Proceedings,* Vol. 663, 2000.

[82] J. Swanson, K. Simmons, D. Norden, and D. Reed, "Microbial Characterization of Halite and Groundwater Samples from the WIPP," Los Alamos National Laboratory, Carlsbad, NM, LA-UR 13-26280, 2013.

[83] R. H. Vreeland, A. F. Piselli Jr., S. McDonnough, and S. S. Meyers, "Distribution and diversity of halophilic bacteria in a subsurface salt formation." *Extremophiles,* Vol. 2, pp. 321-331, 1998.

[84] C. R. Woese, O. Kandler, and M. L. Wheelis, "Towards a natural system of organisms: proposal for the domains Archaea, Bacteria, and Eucarya." *Proceedings of the National Academy of Sciences,* Vol. 87, pp. 4576-4579, 1990.

# Appendix

Tab. XIII Microorganisms under investigation with respect to the degradation of cellulose, bitumen, compacted wastes, and cement as well as species found in the Gorleben cap rocks

| Type | Genus | Species | Cellulose | Bitumen | Waste | Cement | Cap Rock | Notes |
|---|---|---|---|---|---|---|---|---|
| **Bacteria** | *Escherichia* | *coli* | | | | | | |
| | *Pseudomonas* | *fluorescens* | | | x | | | Aerobic |
| | *Brevundimonas* | *vesicularis* | | | | | x | |
| | *Brevundimonas* | *diminuta* | | | | | x | |
| | | *alcaligenes* | | | | | x | |
| | *Delftia* | *acidovorans* | | | | | x | |
| | | *Asphaltenicus* | | x | | | | Organism not found in database, probably renamed |
| | | *aeruginosa* | | x | | | | |
| | | *putida* | | x | | | | |
| | | *aeruginosa* | | x | | | | |
| | *Azotobacter* | | | | x | | | aerobic, free-living soil microbes |
| | *Arthrobacter* | | | x | x | | | soil bacteria can grow on phosphate and sulfate salts |
| | *Bacilli* | *Bacillus cereus* | | | x | | x | aerobic or anaerobic |
| | | *Bacillus circulans* | | | | | x | |
| | *Thiobacillus* | | | | | x | | elementary sulfur, thiosulfate or polythionates as energy sources |
| | *Other bacteria* | *Brevibacterium linens* | | | | | x | NaCl tolerable |
| | | *Staphylococcus* | | | x | | | |

| **Type** | **Genus** | **Species** | **Cellulose** | **Bitumen** | **Waste** | **Cement** | **Cap Rock** | **Notes** |
|---|---|---|---|---|---|---|---|---|
| | | *aureus* | | | | | | |
| | | *Streptomyces griseus* | | | x | | | soil / deep sea |
| | *Halobacterium* | *salinarum* | x | | | | | NaCl tolerable |
| | *Haloarcula* | | x | | | | | |
| | *Haloanaero-bium* | *praevalens* | x | | | | | |
| | *Halomonas* | *Sp.* | x | | | | | |
| | *Halobacter (now Halorhabdus)* | *utahensis* | x | | | | | |
| | *Natronobacterium* | | x | | | | | |
| | *Micrococcus* | *Micrococcus sedentarius* | | | | | x | |
| | *Halococcus* | | x | | | | | |
| | *Clostridium* | *sporogenes* | x | | x | | | anaerobic bacteria |
| | *Nitrosomonas* | | | | | x | | ammonia-oxidizing |
| | *Nitrobacter* | | | | | x | | |
| | *Actinobacterium* | *Mycobacterium* | | | x | | | |
| | *Thermoactinomyces* | | | | x | | | |
| | *Acetobacter* | *pasteurianus* | | | | | x | convert ethanol to acetic acid in the presence of oxygen |
| | *Komagataeibacter* | *hansenii* | | | | | x | |
| | *Corynebacterium* | *lepua* | | x | | | | Exist in asphalt deposits organism not found in database, maybe renamed |
| | *Aeromonas* | *hydrophila* | | | | | x | Exist in brackish water environment |
| | *Suttonella* | *Kingella indologenes* | | | | | x | |
| **mould** | *Aspergillus* | *niger* | | | x | x | | |
| **fungi** | *Mucor* | | | | x | | | |
| | *Rhizopus* | | | | x | | | |

| Type | Genus | Species | Cellulose | Bitumen | Waste | Cement | Cap Rock | Notes |
|---|---|---|---|---|---|---|---|---|
| | *Alternaria* | | | | x | | | |
| | *Cladosporium* | | | | x | | | Halotolerant, isolated from WIPP halite |
| | *Penicillium* | | | | x | | | |
| **yeast** | *Rhodotorula* | | | | x | | | |
| | *Torulopsis holmii* | | | x | | | | |

Tab. XIV Characteristics of the three domains of life [12, 84] .

| | **Bacteria** | **Archaea** | **Eukarya** |
|---|---|---|---|
| Nuclear membrane | No | No | Yes |
| Membrane-bound organelles | Rare, a few types found in a few species | Rare, a few types found in a few species | Multiple distinct types, found in all species |
| Plasma membrane | Similar to Eukarya | Different from Bacteria and Eukarya | Similar to Bacteria |
| Cell wall | Found in nearly all species, constructed of peptidoglycan | Found in nearly all species, constructed of various materials | Found in some species, constructed of various materials |
| RNA polymerases | Single polymerase | Single polymerase, Eukarya-like RNA pol II | Three main polymerases, RNA pol I, II, and III |
| Histones | Histone-like proteins | Yes | Yes, highly alkaline proteins ordering the DNA into structural units |